Machine Learning Techniques for Gait Biometric
Recognition

James Eric Mason · Issa Traoré
Isaac Woungang

Machine Learning Techniques for Gait Biometric Recognition

Using the Ground Reaction Force

 Springer

James Eric Mason
University of Victoria
Victoria, BC
Canada

Isaac Woungang
Ryerson University
Toronto, ON
Canada

Issa Traoré
University of Victoria
Victoria, BC
Canada

ISBN 978-3-319-29086-7   ISBN 978-3-319-29088-1   (eBook)
DOI 10.1007/978-3-319-29088-1

Library of Congress Control Number: 2015960234

Printed on acid-free paper

This Springer imprint is published by SpringerNature
The registered company is Springer International Publishing AG Switzerland

To mom, dad, and my wife Pairin, you helped me get to where I am today. Sombat and Kangwon, oath fulfilled.

To Me-Kon, Mustapha, Khadijah, Kaden, and Ayesha, for your unconditional love and making me a blessed and lucky man.

To Clarisse, Clyde, Lenny, and Kylian, for being there for me all the time. Your endless love, support and encouragement, and push for tenacity, are very much appreciated.

Preface

The last two decades have seen a dramatic increase in the number of stakeholders of biometric technologies. The quality of the technologies has increased due to an improvement in underlying data processing and sensor technologies. A growing and healthy marketplace has emerged, while the number of people using, operating, or impacted by these technologies has been growing exponentially. Several new disruptive technologies have emerged, along with the diversification of the devices and platforms where biometrics are provisioned. The ubiquity of mobile phones and the multiplicity and diversity of sensors available for biometric provisioning (e.g., webcam, fingerprint reader, touchscreen, accelerometer, gyroscope, etc.) is contributing significantly to this dramatic growth of the biometric ecosystem.

Gait biometrics is one of the new technologies that have appeared in the past few decades. Gait biometric technology consists of extracting and measuring unique and distinctive patterns from human locomotion. Different forms of gait biometrics are available based on how the gait information is captured (e.g., video cameras, floor sensor, smartphones, etc.). Gait based on the ground reaction force (GRF) is the most recent form of gait biometric technology, which although lesser known than its counterparts, has shown greater promise in terms of its robustness. GRF is a measure of the force exerted by the ground back on the foot during a footstep.

The GRF-based gait biometric is the central topic of this book. Theoretical and practical underpinnings of the GRF-based gait biometric are presented in detail. The main components and processes involved in developing a GRF-based recognition system are discussed from a theoretical and experimental perspective, by revisiting existing research and introducing new results.

While the central topic of the book is GRF-based gait biometric technology, its backdrop is machine learning. Several machine learning techniques used in the literature for GRF recognition are dissected, contrasted, and investigated experimentally.

The book covers the different dimensions required for developing a GRF-based system: theoretical models, experimental models, and implementation issues. It also covers in detail several machine learning algorithms which can be used broadly for biometric recognition technologies and other similar pattern recognition problems (e.g., speech recognition). This book is intended for researchers, developers, and managers and for students of computer science and engineering, in particular graduate students at the Master's and Ph.D. levels, working or with interest in the aforementioned areas.

The book consists of 11 chapters outlined as follows.

Chapter 1 provides a brief introduction to gait biometrics and outlines the context, objectives, and main contributions of the book.

Chapter 2 gives a high-level overview of machine learning and presents the different factors that define gait biometric technology. A high-level discussion of the considerations and issues underlying the design of a gait-based biometric recognition system is conducted.

Chapter 3 describes the field of gait biometrics and provides a historical overview of work that has been done in the field to date. It goes on to explain where the footstep GRF fits into the field of gait biometrics, and reviews the footstep GRF recognition literature that forms the basis for the research presented in later chapters.

This chapter also presents the experimental setup and introduces the methodology used to achieve the research objectives. It covers the selection of a development dataset containing data of a single shoe type and proposes a biometric system composed of feature extractors, normalizers, and classifiers to perform GRF-based person recognition.

Chapter 4 outlines the purpose and mechanisms underlying feature extraction, and compares four different feature extraction techniques previously used for GRF-based recognition in other studies. Theoretical background is provided for each feature extractor together with a discussion of each implementation. Preliminary GRF recognition results are acquired using the development dataset and presented for the parameter optimization of each extractor.

Chapter 5 demonstrates the performance of various normalization techniques on the extracted feature spaces from Chap. 4. Two novel normalization techniques are introduced here and theoretical background is provided for these and several other well-known existing techniques that are also examined. To determine the effectiveness of normalization the results from applying these normalization techniques are compared with the non-normalized results of the previous chapter.

Chapter 6 presents the theoretical background and implementation for five different classifiers that were selected for analysis in this book. Each classifier is tuned across the best-performing feature spaces acquired from development dataset in Chaps. 4 and 5. Finally, the feature extractor-normalizer-classifier combinations that achieved the best results are summarized for comparison with the results over the evaluation dataset in Chap. 8.

Chapter 7 outlines the experimental method and dataset used to evaluate the proposed GRF-based gait biometric framework. Different evaluation criteria are presented and discussed.

Chapter 8 demonstrates the results obtained after applying the best footstep GRF-recognition systems, outlined in Chap. 6, to an evaluation dataset containing previously unseen data samples with different shoe types.

Chapter 9 discusses the findings behind the GRF footstep recognition experiment. The effects of various techniques are compared, with practical implications and explanations for possible sources of error presented. Finally, the chapter concludes by examining techniques that could potentially be used to improve upon the results discovered in our research.

Chapter 10 identifies various applications for gait biometrics, and discusses current usage of the biometric both commercially and in research.

Chapter 11 provides a final summary of the research presented in previous chapters. The major findings are highlighted and the remaining problems and areas for future work are discussed.

The above chapters present state-of-the-art GRF-based gait biometric research and expose different dimensions of this emerging field, by emphasizing both the theoretical and practical underpinnings. We hope that this will represent a useful resource for readers from academia and industry in pursuing future research and developing new applications.

Acknowledgments

As authors of this book, we want to acknowledge those who helped in its creation.

In particular, we would like to thank Mr. Quin Sandler and Norman Mckay, CEO and Product Architect at Plantiga Technologies Inc., for the fruitful discussions on the industrial perspectives of gait biometric technologies, and helping in our search for useful datasets.

Special thanks should go to the University of Calgary Faculty of Kinesiology laboratory for sharing their GRF datasets.

The research presented in this work was made possible thanks to research grants provided by the Natural Sciences and Engineering Research Council (NSERC) of Canada.

Contents

About the Authors

James Eric Mason obtained his BSEng and MASc from the University of Victoria, Canada, in 2009 and 2014, respectively. During his Master's program, under the supervision of Dr. Issa Traoré, his research focused primarily on biometric security solutions with a particular emphasis on the gait biometric. In 2014 he completed his thesis titled *Examining the impact of Normalization and Footwear on Gait Biometrics Recognition using the Ground Reaction Force*, which served as an inspiration for the work presented in this book. His research interests include biometric security, machine learning, software engineering, web development, and weather/climate sciences. Since 2011, he has been working with the software startup Referral SaaSquatch as a full stack software developer.

Issa Traoré obtained a Ph.D. in Software Engineering in 1998 from Institute Nationale Polytechnique (INPT)-LAAS/CNRS, Toulouse, France. He has been with the faculty of the Department of Electrical and Computer Engineering of the University of Victoria since 1999. He is currently a Full Professor and the Coordinator of the Information Security and Object Technology (ISOT) Lab (http://www.isot.ece.uvic.ca) at the University of Victoria. His research interests include biometrics technologies, computer intrusion detection, network forensics, software security, and software quality engineering. He is currently serving as Associate Editor for the International Journal of Communication Systems (IJCS) and the International Journal of Communication Networks and Distributed Systems (IJCNDS). Dr. Traoré is also a co-founder and Chief Scientist of Plurilock Security Solutions Inc. (http://www.plurilock.com), a network security company which provides innovative authentication technologies, and is one of the pioneers in bringing behavioral biometric authentication products to the market.

Isaac Woungang received his M.Sc. and Ph.D. degrees, all in Mathematics, from the University of Aix Marseille II, France, and University of South, Toulon and Var, France, in 1990 and 1994 respectively. In 1999, he received an M.Sc. degree from the INRS-Materials and Telecommunications, University of Quebec, Montreal, QC, Canada. From 1999 to 2002, he worked as a software engineer at Nortel Networks, Ottawa, Canada, in the Photonic Line Systems Group. Since

2002, he has been with Ryerson University, where he is now a full professor of Computer Science and Director of the Distributed Applications and Broadband (DABNEL) Lab (http://www.scs.ryerson.ca/iwoungan). His current research interests include radio resource management in next-generation wireless networks, biometrics technologies, network security. Dr. Woungang has published 8 books and over 89 refereed technical articles in scholarly international journals and proceedings of international conferences. He has served as Associate Editor of the Computers and Electrical Engineering (Elsevier), and the International Journal of Communication Systems (Wiley). He has Guest Edited several Special Issues with various reputed journals such as IET Information Security, Mathematical and Computer Modeling (Elsevier), Computer Communications (Elsevier), Computers and Electrical Engineering (Elsevier), and Telecommunication Systems (Springer). Since January 2012, he is serving as Chair of Computer Chapter, IEEE Toronto Section.

Abbreviations

AFIS	Automated fingerprint identification system
AIC	Akaike Information Criterion
ANCOVA	Analysis of covariance
ANN	Artificial neural network
ATM	Automated teller machine
CNN-NDEKF	Cascade neural network with a node decoupled extended Kalman filtering
Coif	Coiflet
CRM	Cross cyclical rotation metric
CS	Center star
DARPA	Defense Advanced Research Projects Agency
Daub	Daubechies
DET	Detection error tradeoff
DFT	Discrete fourier transform
DTW	Dynamic time warping
EER	Equal error rate
EMFi	Electromechanical film
ENN	Euclidean nearest neighbor
FAR	False acceptance rate
FDA	Fisher discriminant analysis
FFT	Fast fourier transform
FRR	False rejection rate
FS	Floor sensor
FTA	Failure to acquire
FTE	Failure to enroll
GFSVM	Genetic fuzzy support vector machine
GPCA	Generalized principal component analysis
GRF	Ground reaction force
HCI	Human–computer interaction
HMM	Hidden Markov model
ISO	International organization for standardization

KNN	K-nearest neighbors
KPCA	Kernel principal component analysis
KUDA	Kernel uncorrelated discriminant analysis
LDA	Linear discriminant analysis
Lege	Legendre
LLSR	Localized least squares regression
LLSRDTW	Localized least squares regression with dynamic time warping
LOOCV	Leave-one-out-cross-validation
LSPC	Least squares probabilistic classification
LTN	Linear time normalization
LVQ	Learning vector quantization
M^3-net	Max-Margin Markov network
ML	Maximum likelihood
MLP	Multilayer perceptron
MV	Machine vision
PC	Principal component
PCA	Principal component analysis
pHMM	population hidden Markov model
PIN	Personal identification number
PLS	Partial least squares
PSD	Power spectral density
QDA	Quadratic discriminant analysis
RBF	Radial basis function
RLDA	Regularized linear discriminant analysis
ROC	Receiver operating characteristics
SSS	Small sample size
SVD	Singular value decomposition
SVM	Support vector machine
ULDA	Uncorrelated linear discriminant analysis
WP	Wavelet packet
WPD	Wavelet packet decomposition
WS	Wearable sensor

List of Figures

List of Tables

Chapter 1
Introduction to Gait Biometrics

Over the past several decades national security concerns and the need to deter increasingly sophisticated fraudsters have driven the demand for a new generation of reliable person identification tools. Traditional identification technologies have been built around *something a person has* (such as an identification card) or *something a person knows* (such as a password), but to improve reliability, newer technologies are increasingly including *something a person is*, the physical and behavioral characteristics that define an individual. As technology continues to improve, the automatic recognition of a person based on physical or behavioral characteristics, referred to as biometric recognition, seems destined to have a profound impact on physical and cyber security while we progress through the twenty-first century.

1.1 Context

The first automated biometric system was a fingerprint identification tool developed in the 1970s. This tool, called Automated Fingerprint Identification System (AFIS), was used to assist with forensics investigations of criminal activities. Prior to the mid-1990s biometric devices were typically bulky and expensive, making them difficult to deploy; but with the recent rapid expansion in computing power, it has become much easier to deploy biometric systems. The decreasing cost and size of biometric devices has now made it practical to install them for instant identification at everyday access points, whereas formerly, these devices were reserved for law enforcement or high security environments. However, while technology has enabled a wider use of biometrics, it has also made it easier to circumvent them.

Well-known biometrics based on physical characteristics, including fingerprints, facial features, and iris patterns have shown vulnerabilities to spoofing attacks. The paper, "Biometric attack vectors and defences" [9], by Chris Roberts, referenced a number of successful attacks targeting physical biometrics over the past 15 years. It was discovered that fake fingerprints made from gelatin, and taken from enrolled persons, were able to fool optical fingerprint devices with false acceptance rates as high as 68–100 %. Even more alarming, one team of researchers discovered a technique to successfully "lift" residual fingerprints from scanners using graphite

© Springer International Publishing Switzerland 2016
J.E. Mason et al., *Machine Learning Techniques for Gait Biometric Recognition*,
DOI 10.1007/978-3-319-29088-1_1

powder, tape, and enhanced digital photography, opening the possibility for easy access to sensitive biometric data. Meanwhile, facial recognition has been found vulnerable to spoofing attacks that involved playing back images of a person's face. Furthermore, iris scans have also been successfully spoofed, using high resolution photographs of an enrollee's iris.

To address the potential for spoofing, Roberts [9] suggested several techniques, with a primary focus on the increasing complexity of the data collection process and the capturing proof that an incoming data came from a living person. Such techniques include: requiring blinking, randomization of fingers asked for during a fingerprint scan, thermal measurements, and surface reflectivity among others. There is another category of biometrics for which a living person is often considered as an implicit part of data. This category of biometrics is known as *behavioral biometrics*, and refers to the measurable characteristics of a person's actions. The strength of these biometrics comes from their dynamic nature (the relative ease of requiring variability during identification) and the complexity required to reproduce, and thus spoof them, as well as the actions observed. Such biometrics commonly include speech recognition, keystroke dynamics, mouse dynamics, and walking gait. Although recognition performance by behavioral biometrics is typically weaker than physical biometrics, this category of biometrics presents a major advantage regarding user acceptance, as they are often seen by people to be less intrusive than physical biometrics [13].

Despite behavioral biometrics being applied very early on in the human history for identification in the form of written signatures, it was not until the advent of computers that these biometrics achieved the reliability and automation needed for most practical applications. In the 1960s and 1970s, researchers embarked on the first work aiming to automate behavioral biometrics recognition, initially focusing on written signature and speech recognition [2, 3]. More recently, the human–computer interaction (HCI) biometric has received increasing attention as researchers have come to understand that humans leave digital signatures through their interactions with keyboards and other inputs, as well as in their typical activities performed while on the computer, behaviors that may be uniquely identifiable [14]. Another biometric, which historically was analyzed only in the context of medical science, but in recent years has come to be seen as a potentially useful characteristic to include in biometric analysis, is the *gait biometric*.

The gait biometric reflects the characteristics of human locomotion that can be used to uniquely differentiate one individual from another. Multiple physical factors contribute to the human gait including height, weight, leg lengths, and joint proportions among others, and it was suggested as early as 1967 in a biomedical study by Murray et al. [6] that, together, these individual factors form a pattern of gait unique to every individual. More recent work involving identical twins, which have long plagued other "at a distance" biometrics like facial recognition [4], has shown promise as researchers have sought to verify Murray's suggestion of gait uniqueness. In addition, the fact that gait represents one of the few biometrics that can be captured at a distance attracted the attention of the Defense Advanced Research Projects Agency (DARPA), which included gait along with facial recognition in its Human Identification at a Distance program [8]. This development led to the creation of a publicly available database containing visual gait samples, which has helped to support an increased level of research interest in utilizing the gait biometric.

On inspection, the gait biometric offers a number of advantages that would make it a desirable choice for implementation, where biometric recognition or identification is required. Such advantages include the aforementioned ability to be detected at a distance, the ability to be observed in an unobtrusive manner, the fact that it may be implemented using relatively inexpensive technologies, and the implicit difficulty that would be required to reproduce another person's gait to spoof a biometric system. However, in spite of these capabilities, the gait biometric is also subject to a number of *limitations* that have hindered its incorporation for practical purposes. From a technical perspective, the gait biometric may be obscured by changes in clothing, backpacks, and shoe type, among other factors; it is also subject to the condition of the person being observed and factors such as injury or intoxication may negatively impact its performance. Moreover, from the perspective of *privacy*, certain gait monitoring techniques like video capture have encountered objections. To address these limitations, the methods used to capture the gait biometric have been *diversified* and some newer approaches have been proposed that examine the characteristics that are less likely to raise privacy objections. There has also been a focus on improving the performance in order to mitigate obstructing factors since research has shifted toward the study of new and powerful machine learning techniques.

1.2 Objectives

The ever-increasing use of biometrics to enhance traditional security devices has come under increased scrutiny in recent years. Privacy advocates often make the argument that biometrics present a high risk in the case of a compromise, as most of the existing technologies cannot simply be reset like more traditional identification mechanisms. Researchers have demonstrated that today's biometric tools may not necessarily be as secure as we might imagine while many end users have shown resistance to the intrusive nature of biometric collection techniques, particularly those involving captured images. Biometrics structured around unique *behavioral*, rather than physical, characteristics have been suggested as a means to provide enhanced security with less risk and greater convenience to end users. Furthermore, as was noted in the previous section, one such biometric factor that has attracted a lot of attention in recent years is the human *gait*.

Much of the recent research into gait biometrics has focused on extracting features from gait sequences captured via video. However, this approach raises similar concerns to those of physical biometrics over both *intrusiveness* and potential for *forgery* (via video playback attack). Moreover, while providing the advantage of being observable at a distance, the captured video form of gait biometric is very susceptible to environmental variance. An alternative gait biometric approach that may be less objectionable and perhaps even more secure, albeit at the cost of measurement from a distance, involves extracting unique walking features from the *force signatures* generated as a person steps over floor plate sensors or sensor-loaded shoes. This footstep-based technique offers several potential advantages over the video approach: it does not require the capture of intrusive images; it is less susceptible to interference from obstructions (i.e., changes in lighting or objects obstructing the view); and its interface requires a complicated transfer of force over a short period of time that, with today's technology, would be very difficult to reproduce in a forgery attack. Nevertheless, this technique is still young and has only been studied by a small number of researchers [10].

Previous attempts at performing footstep recognition have generally focused on comparing the recognition ability of well-known *classifiers* (the *machine learning*-based models that determine the likelihood of an identification match) and/or the discriminative properties of footstep *feature extraction* approaches (machine learning techniques that simplify the amount of work that classifiers are required to do, for instance, by identifying the most discriminative data characteristics). Unfortunately, there is not yet any standard publically available footstep force signature datasets, and, consequently, the studies behind these previous attempts at footstep recognition used different datasets of varying quality, making it difficult to accurately assess the effectiveness of their chosen methods [11]. Moreover, large research gaps remain regarding both the effect of data *normalization* on classification success and that of *shoe type* variation on the recognition performance.

This book aims at presenting gait biometric recognition from the unique perspective of the *footstep force signature*, differing from previous books, including the popular work of Nixon et al. [7] titled "Human Identification Based on Gait", which focused primarily on the component of gait captured via video. The force metric examined by this book is referred to as the *Ground Reaction Force* (GRF) and is a measure of the force exerted by the ground back on the foot during a footstep. The nature of the data acquired via this metric lends itself to machine learning analysis techniques traditionally used in speech recognition, as opposed to those often used in image-based gait analysis, and much of the research presented in this book could easily be extended to other similar domains. The primary objective of this book is the establishment of a methodology by which machine learning techniques can be used to isolate and utilize the unique characteristics that may make the GRF-based gait biometric suitable for human recognition. Moreover, as a secondary objective, this book aims to improve the understanding of the gait biometric and expand upon the work of previous researchers by addressing the aforementioned research gaps. To accomplish these objectives, this book presents a demonstrative experiment that takes a step-by-step approach to the development and analysis of a GRF-based gait biometric system.

The presented demonstrative experiment will help verify two assertions suggested in the preliminary research; namely (1) the *variations in shoe type will have a negative impact on the recognition performance* [1], and (2) a *relationship exists between stepping speed and force amplitude that could possibly be used to improve the recognition performance* via *normalization* [12]. In doing so, this experiment will also present *a comprehensive comparison of existing and novel GRF-based gait recognition techniques,* never before directly compared on a single high quality dataset, to better assess their effectiveness. It is hoped that the research presented in this book will make a significant contribution to the present-day understanding of GRF-based gait recognition and pave the way for the deployment of such technology in a real-world system.

1.3   Summary of Contributions

The contributions of the research presented in this book can be described in the following points:

1.3.1   Feature Extraction

The research presented in this book contributes to the present-day knowledge base for GRF-based gait biometric feature extraction in two ways. Unlike previous studies that extracted feature sets from data obtained using at most three GRF sensor signals, this study extracts a feature set from data obtained using *eight GRF sensor signals*. Additionally, the research presented in this book compares the feature extraction techniques applied by previous studies across different datasets on a single dataset to more accurately assess their relative effectiveness.

1.3.2   Normalization

There has been little-to-no previous research into the effects of *feature set normalization* on GRF-based gait recognition. The research presented in this book contributes to the present-day knowledge base by providing a detailed analysis of normalization based on *stepping speed*. No known previous research has provided such an analysis with regard to the impact of stepping speed as a means to normalize the footstep GRF features. We introduce a novel regression-based approach to the stepping speed-based feature set normalization and compare it with the amplitude-based normalization [10] and the stepping speed-based resampling normalization [5] techniques used in the previous GRF-based gait recognition studies.

1.3.3   Classification

Existing studies have deployed some of the strongest known classification techniques to perform the GRF-based gait recognition. The work presented in this book compares the best of these techniques using features obtained from the novel, normalized, eight-signal feature set discussed in the previous two research contributions. The research presented in this book also contributes to the present-day knowledge base by performing classification using a classification technique that was never before used in the study of GRF-based gait recognition.

1.3.4   Shoe Variation

The final contribution that this book makes to present-day research relates to *variation in shoe type*, which might be expected to affect a system performing GRF-based gait recognition. To date, only a single study [1] has attempted to assess the impact of differences between shoe types used to train a recognition system and those used to authenticate using a recognition system. The research presented here expands on that study, performing a detailed analysis of the recognition results obtained from a dataset containing *three different shoe types*.

References

1. Cattin, Philippe C. 2002. Biometric authentication system using human gait. Ph.D Thesis, Swiss Federal Institute of Technology, Zurich, Switzerland.
2. de Leeuw, Karl, and Jan Bergstra (eds.). 2007. *The History of Information Security—A Comprehensive Handbook*, 1st ed. Elsevier Science.
3. Juang, Biing-Hwang, and Lawrence R. Rabiner. 2005. *Automatic speech recognition—A brief history of the technology development*, 2nd ed. Elsevier Encyclopedia of Language and Linguistics.
4. Mohd-Isa, Wan-Noorshahida, Junaidi Abdullah, and Chikkanan Eswaran. 2012. Classification of gait biometric on identical twins. *Journal of Advanced Computer Science Technology Research* 2(4): 166–175.
5. Moustakidis, Serafeim P., John B. Theocharis, and Giannis Giakas. 2008. Subject recognition based on ground reaction force measurements of gait signals. *IEEE Transactions on Systems, Man, and Cybernetics-Part B: Cybernetics* 38(6): 1476–1485.
6. Murray, Mary Patricia. 1967. Gait as a total pattern of movement. *American Journal of Physical Medicine* 46(1): 290–332.
7. Nixon, Mark S., Tieniu Tan, and Rama Chellappa. 2006. *Human Identification Based on Gait*. New York: Springer.
8. Nixon, Mark S., and John N. Carter. 2004. Automatic gait recognition for human Id at a distance, University of Southhampton, London. Technical Report 2004.
9. Roberts, Chris, 2007. Biometric attack vectors and defenses. *Computers & Security* 26(1): 14–25.
10. Rodríguez, Rubén Vera, Nicholas W.D. Evans, Richard P. Lewis, Benoit Fauve, and John S. D. Mason. 2007. An experimental study on the feasibility of footsteps as a biometric. In *15th European Signal Processing Conference (EUSIPCO 2007)*, 748–752, Poznan.
11. Rodríguez, Rubén Vera, John S.D. Mason, and Nicholas W.D. Evans. 2008. Footstep recognition for a smart home environment. *International Journal of Smart Home* 2(2): 95–110.

12. Taylor, Amanda J., Hylton B. Menz, and Anne-Maree Keenan. 2004. The influence of walking speed on plantar pressure measurements using the two-step gait initiation protocol. *The Foot* 14(1): 49–55.
13. Yampolskiy, Roman V. 2008. Behavioral modeling: an overview. *American Journal of Applied Sciences* 5(5): 496–503.
14. Yampolskiy, Roman V., and Venu Govindaraju. 2007. Direct and indirect human computer interaction based biometrics. *Journal of Computers* 2(10): 76–88.

Chapter 2
Gait Biometric Recognition

The gait biometric has demonstrated potential promises as an alternative or complementary identifier for use in human recognition systems. However, there is no single measure that encompasses the full set of complex dynamics reflecting what we consider to be the human gait. Instead, important aspects of gait can be measured using one or more of several analysis techniques. Among these techniques are *visual approaches* involving cameras, which can capture differing angles of gait from a distance, and *sensor approaches*, which collect information about gait while in contact with the subject being analyzed. In this chapter, we explore the ways in which these varying approaches have previously been applied to achieve gait biometric recognition, while also highlighting important possible areas of concern in their usage with respect to practicality, privacy, and security. This chapter provides a foundation of knowledge with respect to *gait* and *machine learning*, which will be built upon through the remainder of the book as we demonstrate, via the application of powerful machine learning techniques and the levels of gait recognition performance we might hope to achieve using a sensor-based approach for demonstration.

2.1 Introduction to Machine Learning

Data representations of the gait biometric tend to be large and contain distinguishing characteristics that might not be obvious to even an expert; however, the field of *machine learning* provides multiple concepts for dealing with such complexity. In this section, we explore some of these concepts.

2.1.1 Machine Learning Paradigm

Building intelligent systems that can automatically learn from data or make predictions on data is essential for many types of modern applications. At the core of

© Springer International Publishing Switzerland 2016
J.E. Mason et al., *Machine Learning Techniques for Gait Biometric Recognition*,
DOI 10.1007/978-3-319-29088-1_2

the design of such systems is the idea of "learning," [38] which consists of getting the system to accomplish one or more tasks using some intelligent decisions based on data and patterns (direct experience) or some kind of observations, rules, data experienced in the past. For instance, a learning system can be embedded in a robot, allowing it to learn how to move through its environment while avoiding collisions with other objects in its presence. In healthcare monitoring, using a learning system to understand and classify human physical activities such as sitting, lying, walking, running, climbing, etc., can help in assessing the biomechanical and physiological variables of a patient on a long-term basis. Other practical examples justifying the need for building intelligent systems that are capable of learning include speech recognition, image and text retrieval, commercial software, and web page classi-fication, to name a few. Learning systems can be built using artificial intelligence techniques such as data mining, learning automata [38], and machine learning [35], to name a few. The latter is considered to be a future driver of innovation [34].

Machine learning [35] is a paradigm that focuses on devising computer pro-grams (often called *agents*) that are able to learn from present and/or past experi-ence or adapt to changes in environment without human user assistance. Its fundamental goal is to *generalize* beyond the examples in a training set. These learning algorithms must satisfy certain requirements in terms of *practicality* and *efficiency*, depending on their learning style and the way they model the problem [11]. They must also rely on the so-called *learning models*, themselves distin-guishable by the type of learning that they induce, which include supervised learning, unsupervised learning, reinforcement learning, and deep learning, to name a few. From a design perspective, these learning options differ from each other in the way they capture the most important learning features, usually, under some simplifying assumptions. These learning features are concerned with determining which patterns are subject to learning, finding how these patterns are generated and assigned to the learning algorithm, and determining the expected learning goal(s).

2.1.2 Machine Learning Design Cycle

The use of a machine learning approach relies on the application of a machine learning design cycle, which consists of a *dataset and feature selection*, a *learning algorithm*, and an *evaluation module*, as shown in Fig. 2.1.

Dataset: This is the input to the machine learning method, which may or may not require a *preprocessing* step (also called preparation step). The preprocessing step is meant to avoid incompleteness or inconsistency in data, for instance, due to missing attributes or attribute values, discrepancy in coding or naming conventions, errors, noisy data, or outliers, to name a few. It is also meant to check for *data biases*, i.e., the data on which the conclusions rely should be similar to the ones that were used for analysis.

The data preprocessing step can be realized using any of the following methods or a combination of them:

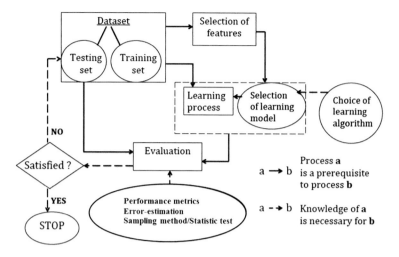

Fig. 2.1 Machine Learning Design Cycle. This figure demonstrates a typical machine learning design cycle

- *Data cleaning* [15]—This process is meant to identify incomplete, inaccurate, and unreasonable data and to improve the quality of data by correcting the detected errors, omissions, outliers (geographic, statistical, temporal or environmental). Its goal is to ensure some conformance to a set of predefined rules. Data cleaning methods include filling out the missing attribute values, for instance, by predicting them using a learning algorithm such as decision tree, binning the attribute values, i.e., transforming continuous values to a finite set of discrete values, clustering the attribute values, i.e., grouping the attribute values in clusters, removing redundancies, performing format checks, completeness checks, reasonableness checks, and validation checks, to name a few.
- *Data normalization* (also called *rescaling*)—This consists of scaling the attribute values so that they belong to a specific prescribed range, e.g.: [0,1] or [−1,1]. Typically, there is no hard requirement for data normalization as preprocessing is more dependent on the data than on the learning algorithm. For instance, for neural networks or deep learning-based learning algorithms where a gradient method is used along with the same learning rate across all weights, all input dimensions can be normalized so that they share a similar distribution.
- *Data aggregation*—This consists of presenting the attribute values in a summarized format; for instance, in the form of their minimum value, maximum value, or average value. Typically, a data cube is constructed for the analysis of the data at various granularities.
- *Data abstraction/new data construction*—This is meant to merge, add, or replace new attributes/attribute values.
- *Data reduction*—This consists of using partitioning, compression, and discretization techniques to reduce the volume of data while guaranteeing similar analytical results and without compromising the integrity of the original data.

Features Selection: As the size of the dataset can turn out to be very large, it is essential to investigate its quality in terms of irrelevant and redundant information, prior to applying the learning algorithm. Feature selection is a method used to discover and remove such information (as much as possible), with the goal of constructing a small subset of highly predictive features so as to avoid *overfitting* the training data. Common feature selection methods include:

– *Wrapper methods* [6]—which rely on the use of a random statistical resampling technique to determine how accurate the feature subsets will be. These methods typically require that the learning algorithm be repeated several times using a different set of features; hence, they may be computationally expensive;
– *Filter methods* [6]—which are used to identify and remove the undesirable features of the data without the need for a learning algorithm. These undesirable features can be identified by examining the consistency in the data or by finding some correlations between them, or by other means, usually based on heuristic search methods that are designed according to the general characteristics of the dataset. As an example, features can be ranked based on some ranking criteria and the top ranking features can be selected.

In a nutshell, a feature selection method typically involves a generation procedure that produces some subsets of features of the dataset, which are then fed into an *evaluation function*[1] that iteratively works for determining the goodness value of each of these subsets by comparing it against the previous best such value based on a predefined stopping criterion. Next, a validation step is required in order to check the validity of any generated subset. In most cases, the efficiency of the feature selection method [17] is measured by either the level at which it allows the chosen learning algorithm to operate in a fast and effective way, or at which it allows for a simple and comprehensive representation of the future classification and its accuracy. A comprehensive taxonomy of feature selection algorithms is provided in [48].

Model Selection: This step refers to choosing a model (or a set of models) from a set of available ones that has the best possible inductive bias given a finite set of training data[2] and the above-mentioned small subset of highly predictive features derived from the *Feature Selection* step. A learning algorithm should also be selected. Model selection methods often include the use of data mining approaches and a search strategy, along with a set of criteria used to compare the available models. An initial data analysis and visualization method may also be required in order to make good guesses about the data distribution form and the shape of the desired learning function. A comprehensive survey of model selection methods is given in [25, 36]. Typically, a fraction of the training data (referred to as validation data) is set aside and the remaining training data is used to train each candidate

---

[1]More description of the evaluation function is provided in the Subsection entitled "Learning Models".

[2]The training set is typically used to train the classifier, whereas the test set is used to estimate the error rate of the trained classifier.

model. Errors are then evaluated on the held-out data and the selected model is the one with the smallest held-out error. Indeed, the chosen model is expected to work best on the training data, and may be based on either the same learning model with different complexities (hyperparameters) or different learning models. Methods to select the validation data include:

- *Holdout method*— the dataset is broken into two subsets: training set and testing set. A function (so-called approximator) is invoked on the training set only in order to predict the output value for the data in the testing set. The resulting mean test set error is then used for evaluating the model.
- *K-fold cross-validation*—the dataset is broken into K disjoint subsets (typically $K = 10$) and the holdout method is invoked K times, each time using one of the K subsets as testing set and the remaining $K-1$ subsets as training set.
- *Leave-one-out cross-validation* (LOOCV)—This is a special case of K-fold Cross-Validation where K is chosen as the total number of examples. For instance, for a dataset with N examples, LOOCV will perform N experiments, each of which uses $N-1$ examples for training purpose and the remaining example for testing purpose.
- *Random subsampling cross-validation*—where: (a) the dataset is randomly split into two subsets of sizes defined by the user (training and test sets); (b) the model is then fit using the training set, then (c) evaluated using the testing set, and (d) the procedure (a)–(c) is repeated multiple times and the mean estimated error is determined.
- *Bootstrapping*—method for which the dataset, say of size N, is sampled with replacement to create a new dataset of the same size before splitting it up into testing set and training set.
- *Other methods*—Model selection can also be thought of from a Bayesian decision–theoretic perspective. For instance, through the design of a statistical analysis technique where the goal is to study how different aspects of the observed units of the dataset affect some outcome of interest. In this regard, information criteria methods [31] can be used as a tool for optimal model selection among a set of available models. The Akaike information criterion (AIC) [2] is an example of an information criterion method that can be used to measure the goodness of fit or uncertainty for the considered range of data values. The minimum description length [31] is another method for model selection, which relies on the principle that any regularity in the data can be exploited to describe it, to name a few.

Learning Process: In this step, the chosen learning algorithm works with a given or modified dataset D of examples—which, for instance, represent some past experiences. It then tries to either describe that dataset in some meaningful way or develop a suitable response for the future data. As an example, in a healthcare environment, an illustration of this scenario is when the learning algorithm is provided a list of patients' records and their corresponding diagnoses; then it tries to learn from this list in order to discover the dependencies between diseases or predict how often the disease will affect future patients with similar type of records.

Typically, a training set of examples, say (x_k, y_k) in $D = \{d_1 = (x_1, y_1), ..., d_n = (x_n, y_n)\}$ is provided by the learner, where $x_k = (x_{k,1}, ..., x_{k,p})$ is an observed input and y_k is the output corresponding to it. The learner then outputs a classifier. The goal of the learning process is to *generalize* beyond the examples in the training set; for instance, to determine whether for future examples x_u the classifier[3] will yield the correct output y_u or will accurately predict it. Since there may be many such functions $f: X \rightarrow Y$ to learn, the question that arises is how to select the best possible function f that will represent the mapping between x_k and y_k? Usually, the learner is expected to embody some assumptions and knowledge beyond the data that it has on hand, for instance, with regard to the type of function to be considered; say for instance $f(x) = ax + b$ for each pair of parameters a and b. This equation defines the learning model M. Here, the parameters a and b should be set in such a way that the misfit between the model M and the observed data D (so-called error or evaluation function) is optimized or the data D is explained in the most meaningful way. Examples of evaluation functions include the *mean square error* and *average misclassification error* [30], to name a few. It should be noticed that the above-mentioned *overfitting* problem occurs when the knowledge and data that the learner has in hand are not sufficient to completely determine the correct classifier.

The choice of the above function f to formally represent a learner is paramount to the choice of the set of classifiers that can be learned (also called *hypothesis space*). This imposes the constraint that only classifiers belonging to this space are considered valid. Next, to differentiate good classifiers from bad ones, the aforementioned *evaluation function* used by the learning algorithm may be different from the one defined in the evaluation step (as shown in Fig. 2.1)— i.e., one that the classifier is expected to optimize. Next, a method to search for the classifier with the highest score among that of all candidate classifiers is needed, which will determine how efficient the learner is. This method relies on the choice of a suitable *optimization* technique. Often, off-the-shell optimizers are first experimented with, and then customized ones are designed later on to replace them. The learning process is therefore an optimization problem that may be difficult to solve because the right choice of the learner and evaluation functions matters.

In a nutshell, a machine learning technique involves three main components: a formal learner representation, an evaluation function, and an optimization technique to find the classifier with the highest possible score:

- *Learner representation*: One can distinguish between *parametric* (or fixed-size) learners—which are useful when enough training samples are provided—and *nonparametric* (or variable size) learners—whose learning efficiency typically depends on whether a sufficient training set is acquired or not, which is difficult to judge due to the *curse of dimensionality*; a factor encountered when the dimensionality (i.e., the number of features) of the examples increases [30].

---

[3]A classifier is defined as a machine learning system that takes, as input, a vector of continuous or discrete feature values and generates a single discrete value (so-called class) as output.

- *Evaluation function*: This is used by the chosen learning algorithm to characterize the goodness of the classifier. This function can be based on performance and/or statistical measures such as accuracy, error rate, precision and recall, squared error, likelihood, posterior probability, information gain, K-L divergence, cost, utility, and margin, to name a few.
- *Optimization technique*: The output is the classifier that has the highest possible score. Continuous constrained optimization such as linear or quadratic programming, unconstrained continuous optimization such as gradient descent, quasi-Newton methods, and conjugate gradient, in addition to combinatorial optimization techniques such as greedy search, branch-and-bound, and beam search techniques, to name a few, can be used for such purpose.

It should be noticed that choosing or defining each of the above three components is a difficult task. Indeed, the fact that a function can be represented does not mean that it can automatically be learned. As an example, a decision tree learner cannot operate on trees that have more leaves than the training set of examples. As another example, in a scenario where the *Features Selection* step has resulted in the construction of the best possible set of features, but the classifiers obtained from the learner are still not accurate enough, more examples and/or raw features must be gathered or a better learning algorithm must be designed. These examples show that several tradeoffs should be considered when designing a classifier or when using an existing one [4, 53].

Various classification algorithms are available in the literature [56]. These may be grouped based on (1) learning style—i.e., the way that the algorithm models the problem based on its interaction with the input data, (2) similarity in terms of function used, and (3) other criterion.

- *Learning style-based classifiers*: These include three main types of learning schemes: supervised learning, unsupervised learning, semi-supervised learning, and reinforcement learning.

  - *Supervised learning*: In this learning model, the training data includes the input and desired responses. The goal is to construct a predictor model that can yield reasonable predictions for the responses to new input data. For some training sets, the correct responses (targets) are known and are given as inputs to the learner during the learning process. In this case, constructing the proper test, training, and validation sets is crucial, and the goal is to obtain correct results when new data are given as input without a priori knowledge of the target.
  - *Unsupervised learning*: In this learning model, the input data is not labeled and the learner is not provided with known results during the training phase. The goal is to have the learner predict the learning relations that are present between the data components.
  - *Semi-supervised learning*: In this learning model, the training data is a mixture of labeled and un-labeled samples and the learner has to learn the structures of the input data and make predictions for new input data.

  – *Reinforcement learning*: In this learning model, samples of the input vector
    x (but not *y* the corresponding response) are provided to the learner. Instead
    of *y*, the learner gets a feedback (reinforcement) from a critic about how
    good the response was. The goal is to select responses that will lead to the
    best possible reinforcement.

• *Similarity-based classifiers*: These include *instance-based learning* methods
  such as *K-Nearest Neighbors* and *Support Vector Machines*; hyperplanes
  methods such as naïve Bayes and logistic regression; graphical models such as
  Bayesian networks and conditional random fields [5]; set of rules-based methods
  such as proposition rules, logic programs, association rule learning; other
  methods include *artificial neural networks*, decision trees, regularization
  methods, Kernel methods, clustering methods, deep learning, *dimensionality
  reduction*, ensemble methods, and learning vector quantization [11], to name a
  few.

Evaluation: As stated earlier, the learned model has to be applied to new data in
order to predict the responses *y* for new input data *x* using the learned function *f(x)*.
This is referred to as *Evaluation step*, which is meant to determine the *effectiveness*
and *efficiency* of the learned *f(x)* (classifier or learning algorithm) on various collected datasets [30] based on a predefined *evaluation* metric and a *confidence*
estimation method. The goal of the evaluation step is to: (1) compare a new learning
algorithm against other classifiers on a given domain or a set of benchmark
domains, or (2) characterize some generic classifiers on some benchmark domains,
or (3) compare various classifiers on a specific domain. Performance metrics of
interest and a framework for the evaluation of learning algorithms and classifiers are
provided in [30]. To test the learner's performance, the basic experimental setup
consists of: (1) taking the dataset *D* and dividing it into training dataset and testing
dataset (for instance, using the above-described model selection methods), (2) using
the training dataset and the chosen learning algorithm to train the learner, and
(3) testing the learner on the testing dataset. The results obtained with the testing
dataset can then be used for comparing other learners with different models and
learning algorithms.

2.2 General Principles of Designing Gait Biometric-Based Systems

The machine learning concepts from the previous section outlined some of the
potential building blocks for creating a gait recognition tool. To make use of these
concepts and exploit the gait biometric for practical purposes, we must first design a
biometric system that is suitable for the requirements of the intended usage. The first
question that should be asked when designing a biometric system is whether the
system will be used for *identification* or *verification* purposes. If a system is

configured for verification, it will be provided a claimed identity information alongside a subject's gait biometric information and will be responsible for deciding whether that identity matches the provided biometric information; such systems are typically used for user authentication. On the other hand, if a system is configured for identification, it will only be provided a gait biometric sample and will be responsible for assigning an *identity* to the sample; such systems are typically used in forensics analysis, surveillance, and catering (i.e., they provide a custom experience based on an individual's preferences). The choice of system together with its intended function may impact the design decisions regarding the system's sensitivity to errors.

When designing a gait biometric system for the purpose of verification, also referred to as the verification mode of operation, there are two key metrics that must be addressed: the *False Acceptance Rate* (FAR) and *False Rejection Rate* (FRR). The FAR (obtained via Eq. 2.2) is a measure of the rate at which the verification attempts are incorrectly accepted, reflecting the number of samples incorrectly accepted (N_{FA}) as a proportion of the total number of samples accepted (N_{SA}); a high FAR would indicate that the system is vulnerable to intrusions. Conversely, the FRR given in Eq. (2.2) is a measure of the rate at which the verification attempts are incorrectly rejected, reflecting the number of samples incorrectly rejected (N_{FR}) as a proportion of the total number of samples rejected (N_{SR}); a high FRR would indicate that the system is prone to locking out its intended users, and as such has a low usability. In the verification mode, the biometric system's machine learning algorithms will typically produce a percentage output reflecting a certainty as to whether the subject being verified matches a provided sample, and, from this, a threshold can be chosen at which a given subject will be accepted or rejected by the system.

$$FAR = \frac{N_{FA}}{N_{SA}} \times 100 \tag{2.1}$$

$$FRR = \frac{N_{FR}}{N_{SR}} \times 100 \tag{2.2}$$

In general, the goal of the biometric system in verification mode will be to simultaneously reduce both error rates. This goal can be accomplished by designing the biometric system in a way that minimizes a third metric referred to as the *Equal Error Rate* (EER). The EER identifies the single rate at which a biometric system's FAR is equal to its FRR, when given some input dataset. This metric can be visually observed using a *Detection Error Tradeoff* (DET) curve, as shown in Fig. 2.2.

Designing a gait biometric system using the identification mode of operation presents a slightly different set of concerns. This type of system may be designed using one of the following two approaches: a "closed set" approach whereby all subjects being identified are known to the system or an "open set" approach whereby subjects unknown to the system might be presented for identification. If only a small group of known users will ever be put up for identification, then the closed set approach might simplify the design of a gait-based biometric system. Moreover,

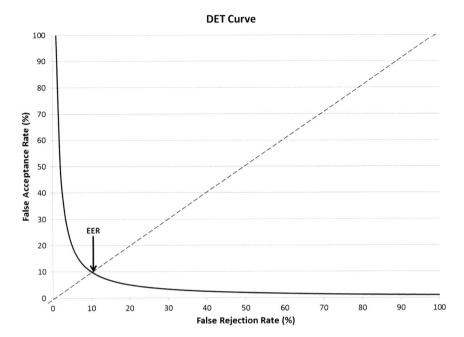

Fig. 2.2 Example DET Curve. This figure demonstrates an example of a DET Curve with an EER around 10 %

under a closed set system, the design will be faster and easier to evaluate, since it is assessed using only a single metric, measuring the percentage of correct identification matches, referred to as the identification rate. However, in many real-world identification scenarios, no assumptions can be made about the identity of the people being examined and, consequently, the open set identification system lends itself to more practical applications than its closed counterpart. This open set identification mode is measured according to two primary metrics: the *false alarm rate*, i.e. the percentage of identification instances for which an unknown user is identified as a known user or known user identified incorrectly, and the *detection and identification rate*, i.e. the percentage rate at which known users are correctly identified. In a system using the open set identification mode, the tradeoff between the two aforementioned rates is typically represented using a *Receiver Operating Characteristics* (ROC) curve, as demonstrated in Fig. 2.3. Using this information, it is left to the designer to determine whether the biometric system should be configured to tolerate a higher false alarm rate or lower detection and identification rate.

Having decided upon the mode of operation for a gait-based biometric system, the next consideration should involve the method of capture. The gait biometric is typically measured using one of the following two different approaches. The first approach involves capturing a *video* of a person's walking motion, while the second approach involves *sensors* that are in direct contact with the subject in motion. Each approach carries a unique set of capabilities and limitations that may influence their

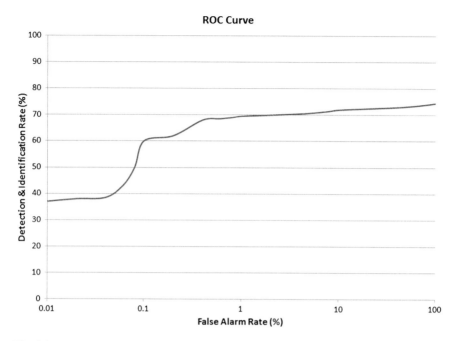

Fig. 2.3 Example ROC Curve. This figure demonstrates an example of an ROC curve used to assess the error tradeoff for a biometric system using an open set identification mode

incorporation into a security system. A video imaging-based capture of the gait biometric may be appropriate in situations where subjects are likely to be uncooperative or must otherwise have their information captured from a distance; it may, however invoke *privacy concerns*. Conversely, the sensor-based approach may allay privacy concerns, but requires a reasonable degree of *cooperation* on behalf of the subject. At other times, the choice of gait capture may be limited by the resources available. Yet, when possible, it may be beneficial to use more than one form of gait biometric captures, as this has been shown to the performance in previous research [13]. The example in Fig. 2.4 demonstrates the three most commonly studied methods of gait capture for biometric identification and verification, including two sensor-based approaches and one image-based approach. The chosen method of gait capture directly affects the format of data samples produced, and therefore will have a key role in the selection of machine learning algorithms most suited for the biometric system design.

At a high level, the *recognition system* from Fig. 2.4 possesses the "intelligence" of the gait biometric system. The design for this component follows a similar pattern in all biometric systems and its interaction with users can be separated into two phases: an *enrollment* phase and a *challenge* (or matching) phase. The diagram in Fig. 2.5 expands upon this component to demonstrate how it facilitates the interactions with the users. In designing a gait biometric system to satisfy this pattern, it is important to consider how the training data will be collected for enrollment, how many

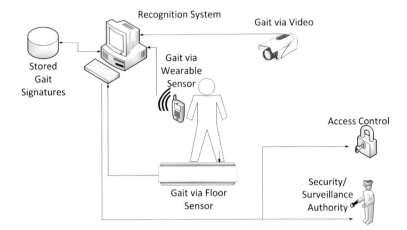

Fig. 2.4 Gait Biometric Capture Techniques. This figure demonstrates an integration of the three most commonly studied gait biometric capture techniques into an example biometric system. In this system the *recognition system* captures sample data from the subject then compares it with previously stored samples or templates (*gait signatures*); the information is then passed on for *access control* or to a *security or surveillance authority* depending on the mode of operation

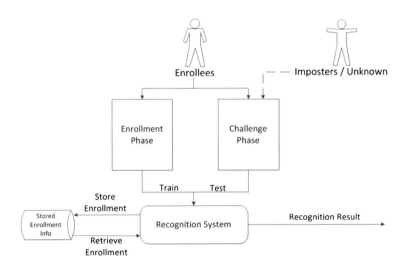

Fig. 2.5 Gait Biometric System Phases. This figure demonstrates how the two phases of user engagement fit into a gait biometric system design. During the *enrollment* phase enrollees would be expected to walk for a certain number of cycles until an adequate amount of information about their gait has been acquired. Later, during the *challenge* phase, the system will use this enrolled data to identify or verify a user while attempting to filter out imposters

samples might be required to sufficiently train the system, and how to ensure security of the stored data. Choices should also be made at this point in the design as to whether the system must produce real-time results, delayed results, or will only be used after the fact; in some cases it may be possible to get more accurate results if real-time data analysis is not a requirement.

Gait data samples tend to be large and contain a high degree of variability, making potentially important patterns within the data difficult to assess visually. Discovering a process that is able to isolate and exploit these patterns to best assist in the distinguishing of one individual from others is at the core of gait recognition-based research. Previous research has suggested that *machine learning* techniques, algorithms that specialize in the extraction and prediction of patterns, are well suited for accomplishing this objective. These techniques have been of particular interest in the areas of feature extraction and classification, but could also potentially be extended to be used in the normalization of data via a form of *regression analysis* (this will be explored in Chap. 5). Before designing a gait-based biometric system, it is important to understand the various forms of machine learning and data analysis techniques that may be included and how they might fit together within a biometric system.

Gait monitoring system inputs typically run as a continuous feed of data, oblivious to the periodic cycles we associate with gait. In video approaches, capturing a cycle might mean starting and stopping the sample extraction cycle when certain movements are detected with respect to a static background, while in sensor approaches, it might mean picking up on force spikes and drop-offs that would be associated with a footstep. Consequently, the first action that will need to be performed in any gait-based biometric system is the *extraction of the sample* over the desired periodic interval. From there, some systems may wish to account for variations in scale, such as the varying distances for which subjects may appear in an image, by performing normalization. Finally, to reduce large sample sizes down to more manageable sizes for classification, a third data preprocessing step, known as *feature extraction*, may be required prior to pushing the sample to the classifier (the final component in all biometric systems). To illustrate the incorporation of these data analysis and machine learning-based components into a gait-based biometric system design, Fig. 2.6 presents the relationship between each as a simple flow diagram.

A final consideration that may arise in the designing of a gait-based biometric system relates to the lack of knowledge about the data that will be collected. If there is no history of biometric systems developed around a chosen data format, then it will be difficult to judge the most suitable values for the machine learning parameters that underpin the system. To optimize the biometric system prior to evaluation, it is generally suggested that the tested data be separated into two mutually exclusive datasets: a *development dataset* and an *evaluation dataset*. Then, using the development dataset, machine learning parameters can be optimized, with the optimal solution being incorporated into the biometric system for an unbiased evaluation via the evaluation dataset; a process we will cover later in this book through our demonstrative experiment.

Now that we have demonstrated the principles on which the design of a gait-based biometric system rely, we can explore some of the ways in which gait

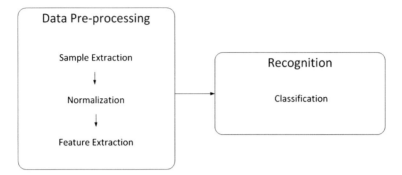

Fig. 2.6 Gait Biometric System Components. This figure demonstrates the breakdown of components that add "intelligence" to a gait-based biometric system in the order in which they would appear in the processing of a data sample. Each of these components may be represented by a number of different data analysis or machine learning techniques, which will be discussed in further detail in later chapters

biometrics have been implemented. The next section provides an overview of previous research as it pertains to the different gait biometrics approaches, together with a brief discussion of *security* and *privacy*-connected issues that relate to such systems.

2.3   Authentication Using the Gait Biometric

Biometric recognition using gait presents a number of advantages over the use of more traditional biometric traits: it is generally considered unobtrusive, as it can be measured in a way that does not require a person to alter his or her typical behavior; it does not require a person to present any more information than is already available to a casual observer; and studies have suggested it is very difficult to imitate [21]. In his research, Cattin [13] (Cattin, 2002) makes special mention of the ability of the gait biometric to perform a *living person test*. The test is described as the ability to determine whether the owner of a trait being observed is alive and physically present or not. Traditional biometrics, such as fingerprints, often fail this test as the traits observed can be faked with present-day technology. The security of the gait biometric lies in the incredible difficulty required to spoof it, thus the living person test is considered intrinsic to the method. Yet, as was mentioned in the previous section, there is no single approach that captures the gait in its entirety. Moreover, while it may be considered less intrusive than other biometrics, it may also be found to be invasive of privacy due to the ease with which it can be used to track individuals without them knowing they are being monitored. In this section, we explore the various approaches to perform authentication via gait biometrics and the impacts that this biometric may have in terms of privacy and security.

2.3.1 Privacy and Security Implications of Gait Biometrics

As gait biometrics slowly find their way into real-world applications, it is worth noting that they carry with them a number of concerns common to *all biometrics* with respect to security and privacy, as well as several concerns more specific to the gait itself. Perhaps most importantly from a security standpoint, gait biometric has not yet achieved the *reliability* that would be required as a standalone biometric, and therefore it is typically used together with more reliable biometrics as part of a *multimodal* biometric system [24]. It has also been shown through the work of Gafurov and Snekkenes [23] that, in the absence of information about enrolled gait users, the gait biometric cannot easily be spoofed; however, with the knowledge that an attacker has a similar gait signature to an enrolled subject, spoofing becomes substantially easier. Such an attack is just one of many that could occur if the attacker were to compromise the non-sensor components that make up the biometric system. Ratha et al. [42] identified eight such potential areas in which a biometric system might be compromised. Of these risks, one of the greatest areas for concern relates to the system's *database*, which if compromised would allow an attacker to view, create, delete, and/or modify the records at will. Another area of concern would be the links between the systems, which if compromised may be subject to replay attacks to gain access to a previous user's template or a bypass of the biometric system altogether in the case of the link granting the recognition decision. To address these concerns, various technical security implementations could be implemented including equipment tampering alarms, encryption of the database, and encryption between links. Moreover, due to the sensitivity of the biometric data, it has been suggested that it only be stored in *nonreversible* template representations rather than as raw data to reduce the utility of the information were it to be compromised [14]. Alternatively, attacks may come in forms that no amount of technical security could ever be expected to prevent, via *social engineering*; an example of such an attack might come as a bribe offered to an employee in return, for access to a system. To address such a case, an organization might wish to conform to the *ISO27001* and *ISO27002* standards for information security [12].

Security in a gait-based biometric system may also be affected by undesirable behavior even without being compromised. One such case would come as a result of this biometric's fluidity and likelihood to change due to environmental factors as well as over the lifetime of an enrolled subject; system training cannot account for all such factors. Consequently, any gait-based security mechanism is likely to eventually start rejecting samples for an enrolled user simply because they do no longer possess the gait dynamics that were present during training; this may restrict the gait from ever being used as a standalone principal biometric. Furthermore, the strength of security in the system will reflect its mode of operation, with identification mode being inherently less secure than verification mode [57]. Nevertheless, a secure incorporation of gait biometrics into a system provides key advantages in enhancing the overall security and making the system less prone to plausible deniability on behalf of a malicious insider.

From a privacy perspective, the gait biometric carries with it some fairly substantial concerns. To begin, the gait biometric like most other biometrics can be considered to be relatively *static* over its intended period of use, and therefore, were it to be leaked, an attacker would gain access to a key piece of information that could not easily be changed or reset like a password or keycard; however, given the dynamic nature and difficultly in reproducing the gait, it is not entirely apparent how useful such information would be to an attacker. A greater potential privacy concern with respect to the gait is the ease with which it could be used to enroll and track individuals *without* their consent or even knowing they are being tracked; such a scenario could play out through an extensive network of video cameras or floor sensors and would violate an individual's right to anonymity. This application of gait biometrics has attracted the greatest amount of attention from privacy advocates, yet, as prominent gait biometric researchers, Boulgouris et al. [9] have pointed out that such a scenario remains technically infeasible for the foreseeable future. A greater immediate privacy concern might come as a result of the use of gait tracking technologies, not for biometric recognition purposes, but as a *surveillance* technique for detecting abnormal behaviors; such behaviors would be flagged as suspicious and may bring scrutiny upon the individual that triggered them. In one recent study, it was demonstrated how gait analysis could detect whether an individual was carrying excess load in addition to its body weight, and evidence suggested it may be possible [46]. If such a tracking system were to be deployed in public, it could be prone to very high false match rates, with undeserved scrutiny being given to individuals suffering from injury, fatigue, or psychological conditions. In a similar fashion, gait analysis could have the dual purpose of identifying the *medical conditions*; for instance, gait analysis has proven particularly valuable in the study of Parkinson's disease [45]. Yet, again, incorporating such technologies into public surveillance systems could compromise the medical privacy rights.

A strong public response should be expected in the event that a gait biometric tracking system was ever modified to take on any of the aforementioned dual-purpose surveillance objectives. Consequently, before deploying a gait biometric recognition system for real-world use, sufficient effort should be made to allay relevant public privacy fears.

2.3.2 Gait Biometric Approaches

There are a variety of characteristics that define the human gait and a variety of techniques used to extract these gait characteristics. In a summary of research in the field of gait biometrics, Derawi et al. [19] have suggested there are 24 different components that, together, can uniquely identify an individual's gait. However, there is no single device able to capture the full complex set of motions that form the human gait. Consequently, the approaches used to accomplish gait recognition relate directly to the instrumentation needed to extract the gait data, and fall into three categories: the *machine vision* approach, the *wearable sensor* approach, and the *floor sensor* approach. The following subsections examine these approaches and describe the research that has been done to date with respect to each.

2.3.2.1 The Machine Vision Approach

The machine vision (MV) approach to gait biometrics typically involves capturing gait information from a distance using video recorder technology. This is the most common approach to gait biometric recognition referenced in the current literature [21], having benefited from the availability of large public datasets such as the NIST MV gait database [40]. Recognition via MV has been accomplished using two different techniques: *model-free* and *model-based* recognition algorithms. The model-free technique, often referred to as the *silhouette*-based technique (Fig. 2.7), involves deriving a human silhouette by separating out a moving person from the static background in each video frame. Using this technique, classifiers are developed around the observed motion of the silhouette. In contrast, the less commonly used model-based technique (Fig. 2.8) involves imposing a model for human movement [39]; this is often accomplished by extracting features, such as limbs and joints, from captured images and mapping them onto the structural components of human models for recognition [16].

Over the past decade gait recognition using MV has been investigated by a number of researchers using a variety of methods with promising results. In 2003, Wang et al. [55] developed a silhouette-based technique that used the feature space dimensionality-reducing principal component analysis (PCA) together with the nearest neighbor and Euclidean nearest neighbor classification algorithms. This approach achieved identification rates in the 70–90 % range, varying on the dataset and acceptance criteria used. Later, in 2005, Boulgouris et al. [10] proposed a novel silhouette-based system for gait recognition using linear time normalization (LTN) on gait cycles. The system demonstrated an 8–20 % improvement in its identification rates when compared with existing methodologies at the time. Another study that same year by Lu et al. [33] achieved an identification rate of 92.5 % using a genetic fuzzy support vector machine (GFSVM) classifier; this

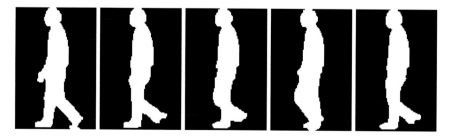

Fig. 2.7 Gait Silhouette Sequence. This figure presents an example of a gait silhouette sequence taken from the publically available OU-ISIR gait samples [29]

Projection of Human Body Model   Modeled Gait

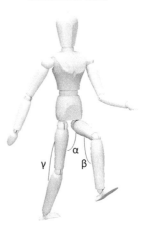

Fig. 2.8 Dynamic Gait Modeling. This figure demonstrates the simple modeling of gait in two and three dimensions. The more advanced *three-dimensional modeling* may require several cameras from different angles but can counter covariate factors like differences in scale that might otherwise lead to invalid results using a silhouette-based MV approach

result improved upon the results of the nearest neighbor and standard support vector machine (SVM) tested against the same dataset.

In 2006 Cheng et al. [16] introduced a gait recognition system that used a hidden Markov model (HMM) and, unlike previous systems, was designed to perform recognition on subjects walking down different paths. It accomplished this by first recognizing the walking direction, then applying an appropriate identifier to the determined path; identification rates achieved by this system varied in the 80–90 % range across differing datasets and acceptance criteria. Another silhouette-based study in 2006, by Liu and Sarkar [32], used a generic population HMM (pHMM) to normalize gait dynamics, then used PCA to reduce the feature space and a linear discriminant analysis (LDA) classifier to perform classification; the use of an HMM for normalization was unique to this study and demonstrated how normalization could be used to improve upon the recognition performance. In 2009, a study by Venkat and De Wilde [54] took a different approach and attempted to reduce the computational intensity of silhouette-based recognition techniques by examining *sub-gaits*, defined as smaller localized frames, rather than entire gait images. This technique yielded an identification rate range of about 75–90 % across various datasets and acceptance criteria. However, when vision or motion obstructing factors such as carrying condition and particularly clothing condition were considered, the identification rate dropped to as low as 29 %.

Few researchers to date have studied the effects of the various factors that can obstruct the human gait recognition; however, a real-world gait recognition system would most likely need to be designed to handle such events. To address the issue of gait variability and obstructions, also known as *covariate factors*, Bouchrika and Nixon [8] proposed a model-based system to extract human joint positions and model gait motion using elliptic Fourier descriptors. Their 2006 study successfully extracted 92.5 % of the heel strikes observed in a dataset containing both visible joints and joints occluded by clothing, demonstrating that the key features of motion analysis could still be tracked in the presence of covariate factors. A follow-up study in 2008 [7] examined the effects of footwear, clothing, carrying condition and walking speed, and, using the model-based approach to joint extraction together with a K-nearest neighbors (KNN) classifier, achieved an identification rate of 73.4 % against a database containing variations of these covariate factors.

Two further studies attempted to mitigate the weaknesses of the MV approach to gait recognition by fusing it with a *secondary* biometric factor. In 2002, Cattin [3] developed a system that fused the data from a video sensor recognition system with a force plate-based footstep recognition system to recognize an individual walking in a monitored room. The results of his study were promising with a verification EER of 1.6 %. In 2006, Zhou and Bhanu [60] developed a different multifactor biometric technique that combined an MV gait recognition system with a facial recognition system. This system had the benefit of only requiring a single video sensor for capturing both biometrics and achieved an identification rate of 91.3 %.

All studies discussed so far have dealt with recognition using images captured from a video sensor; yet one unique study, which, however, was not based on visual data but best falls under the MV category, proposed *audio*-based footstep recognition. In 2006, Itai and Yasukawa [28] took a wavelet transform technique, widely used in feature extraction for speech recognition, and applied it to feature extraction for audible footsteps. Using this technique, an identification rate of 80 % was achieved, suggesting that this could be an exciting new realm for study in the field of gait recognition.

2.3.2.2   The Wearable Sensor Approach

Biometric recognition using wearable sensors (WS) is a new approach to gait biometrics that aims to use sensors attached to the human body (Fig. 2.9) to perform recognition. Much of the early research into gait-aware wearable sensors came from medical studies that focused on their usefulness for detecting pathological conditions [3]. Research into the usefulness of the WS approach for biometric recognition has been, at least partly, held back due to the lack of large publically available datasets [19]. However, the WS approach to biometrics presents a number of advantages, including the ability to perform *continuous authentication*, which would not always be possible with sensors fixed to a physical location. Over the past 10 years, a number of studies, using a variety of techniques, have investigated the feasibility of the WS approach.

In 2006, Gafurov et al. [22] developed a WS biometric recognition system using an accelerometer sensor attached to the lower leg. This study first collected test

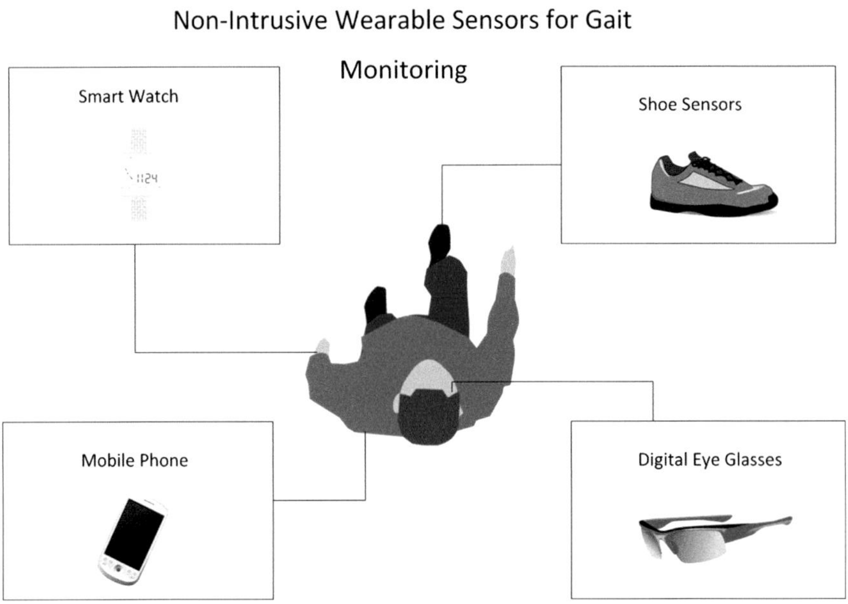

Non-Intrusive Wearable Sensors for Gait Monitoring

Fig. 2.9 Nonintrusive Wearable Sensors for Gait Monitoring. This figure demonstrates some of the nonintrusive devices that can be used to monitor gait via embedded wearable sensors; stronger gait recognition performance may be achievable by combining multiple types of sensor

subject data then uploaded it to a computer, rather than performing real-time classification. The experiment achieved a verification EER of 5 % when recognizing individuals using a histogram similarity classification technique. Another study in 2006, by Huang et al. [27], presented a recognition system based on sensors embedded in a shoe. The sensors used included a pressure sensor, tilt angle sensor, gyroscope, bend sensor, and accelerometer. This system was developed to transmit gait data from the shoe to a computer in real time, and used the PCA feature reduction technique together with an SVM classifier to accomplish a 98 % identification rate on a small sample dataset. A follow-up study by Huang et al. [26] (Huang, Chen, Huang, & Xu, 2007) in 2007 applied a Cascade Neural Network with a node-decoupled extended Kalman filtering (CNN-NDEKF) classifier to the shoe-based WS system and achieved a 97 % identification rate.

One major weakness of the WS approach is the potential inconvenience or discomfort that may be caused by attaching sensors to the human body. For this reason, WS research has tended to focus on one of two unobtrusive WS techniques: *shoe-based* monitoring techniques, like [27] and [26] described in the previous paragraph, and *phone-based* monitoring techniques. Both techniques make use of equipment that is already a part of daily life and require no alterations to typical behavior. In the past few years, the increasing computational power and wider use of smart phones, and increasingly *smart* watches, has sparked a number of studies into feasibility of mobile-based monitoring techniques.

A study in 2009, by Spranger and Zazula [49], described a biometric recognition system that worked with a feature set consisting of cumulants of accelerometer data captured by a mobile phone attached to a person's hip. The system achieved a 93.1 % identification rate on a small dataset using PCA for feature dimensionality reduction and an SVM classifier. A separate study in 2009 by Fitzgerald [20] demonstrated a system, designed for possible future use in mobile phones, that worked with accelerometer and gyroscope data captured by a Nintendo Wii controller. Gait cycles captured by the system were normalized with respect to time. User recognition for the system was examined using KNN, naive Bayes, and quadratic discriminant analysis (QDA) classifiers, with the KNN classifier performing best, achieving an identification accuracy of about 95 %. In 2010, Derawi et al. [18] collected data from accelerometers attached to a belt on the legs of 60 volunteers, generating a much larger dataset than used in the other WS studies described in this chapter. This study focused on cycle length as a metric and, using a cross cyclical rotation metric (CRM), achieved a verification EER of 5.7 % for person recognition. Although the device used by the study was not a mobile phone, the application of this system for use in mobile phones was noted as an important area for future research. In 2011 Nickel et al. used an HMM classifier on accelerometer data from commercially available mobile phones to perform person recognition and achieved a verification EER of about 10 %. The study worked with a relatively large dataset of 48 subjects and was particularly promising for the field of WS-based authentication, because it proved that even a standard, commercially available mobile phone could now be used for gait recognition purposes.

2.3.2.3 The Floor Sensor Approach

The floor sensor (FS) approach to gait biometrics involves recognizing people based on the signals they generate as they walk over sensor-monitored flooring. Data captured by floor sensors typically falls into two categories: *binary image frames* (Fig. 2.11) of the foot while it is in contact with the ground, and single dimensional *force distribution plots* (Fig. 2.10), which describe the force exerted by the foot over time. Most FS technology was developed for the study of biomechanical processes; particularly for improving performance in athletics and discovering the effects of pathological conditions such as diabetes [47]. The first studies using FS technology for gait recognition began in the late 1990s [43]. Over the past 10 years, a small but increasing number of studies have examined the FS approach to gait biometrics. Some, like [13] by Cattin, described in the previous machine vision section, combined the FS recognition approach with another gait recognition approach, such as the MV approach, to improve the recognition accuracy; however, most other studies have focused on using the FS approach for *single factor* recognition.

Open research into using footsteps as a biometric dates back to a 1997 study by Addlesee et al. [1]. In this study, load cell floor sensors were used to capture partial ground reaction force (GRF) data for 15 volunteers and, using an HMM classifier, a

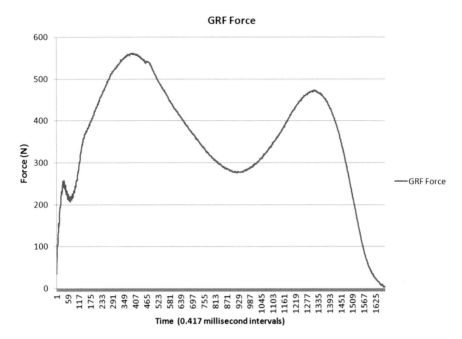

Fig. 2.10 Footstep GRF Force Time Series. This graph demonstrates the force-time series representation of a footstep commonly used for FS-based recognition

91 % footstep identification rate was achieved. Three years later, Orr and Abowd [40] outfitted a floor tile with a set of force sensors to capture the GRF profile for 15 volunteers. In the study, ten features were extracted and normalized, then passed to a Euclidean nearest neighbor (ENN) classifier for recognition; the result was a 93 % identification rate. Another study in 2005, by Suutala and Röning [51] used a floor sensor called ElectroMechanical Film (EMFi) to capture the GRF for ten volunteers. The primary focus of this study was to compare various classifiers, combine such classifiers, and examine the effects of rejecting unreliable data samples from the classifier training. In this study, it was found that the SVM and multilayer perceptron (MLP) neural network classifiers performed best; the strongest corresponding identification accuracy on their most complicated dataset was around 92 %, which increased to 95 % when the most unreliable 9 % of sample set was rejected. A later study in 2007 by Moustakis et al. [37] captured the GRF for a larger dataset of 40 volunteers. This study used a feature extraction technique built on wavelet packet decomposition (WPD) to detect the transient characteristics and distinguishing features of the GRF, then applied an SVM classifier to the feature set; the result was a 98.3 % identification rate.

In 2009, Ye et al. [58] presented a unique technique for FS-based biometrics: instead of performing recognition on a footstep, like most previous studies, they developed a system that could recognize a person by upper-body movements

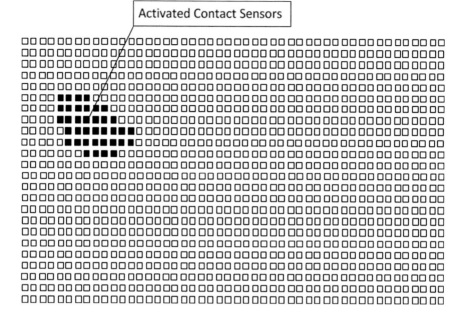

Fig. 2.11 Binary Footstep Frame. This diagram demonstrates a single frame of a footstep on a pressure sensitive sensor array. The darker region represents locations where the footstep is detected as being in contact with the floor. The series of frames produced by a single footstep can be used for FS-based footstep recognition

performed while standing on a force plate. The study obtained the center of pressure for the foot, and monitored its movement as instructed actions were completed by a person on a force plate. This study achieved its best results using a neural network classifier, with a verification EER in the 1-12 % range. Two other studies, one in 2008 [50] and another in 2011 [59], also took a different approach to FS-based gait biometrics, opting to perform gait recognition using binary images of footsteps rather than GRF force signatures. The 2008 study by Suutala et al. [50] examined the shape and pattern of individual footstep image frames, as well as the displacement between the feets during footsteps; it achieved a maximum identification rate of 84 %, using a Gaussian process classifier. The 2011 study by Yun [59] focused primarily on subjects in bare feet and also extracted features from individual footstep frames together with footstep displacement. Using an MLP classifier trained with the extracted features, this study achieved a 96 % identification rate.

Research focused on the FS approach to gait biometrics, like the WS approach, has been disadvantaged due to the lack of any *publically available dataset*. This lack of a publically available dataset makes it difficult to compare results across studies. To address this issue, one group at the University of Wales, Swansea, set out to develop such a dataset. In 2007, Rodríguez et al. [43] published a study of an FS-based recognition system and introduced a dataset of footstep force signatures covering 41 persons and over 3000 footsteps; the intended purpose was to verify the data and make the dataset available at some future point. In their process, they presented a holistic and geometric feature set, and, using an SVM classifier, they achieved a verification EER of 11.5 % for person recognition. A later study [44] in 2008 dealt with a dataset expanded to 55 persons and more than 3500 footsteps; using the same classifier as the previous study, but, additionally normalizing and optimizing the feature set, a verification EER of 13 % was achieved, with the small increase in error rate attributed to the larger dataset. The database in the 2008 study was said to have been made publically available, but, at the time of writing, no longer appears on the project website [52]; nevertheless, the project web site has indicated that a larger dataset is currently being packaged for future release.

2.4   Summary

Like any biometric system, the gait biometric system is primarily a pattern recognition system. As such, it relies heavily on machine learning for human identification and recognition. This chapter gave a brief overview of machine learning and explored the various factors that define biometric recognition via gait. A high-level overview of the complexities that go into designing a system for the purpose of gait-based biometric recognition was provided, and it was shown that there is no single measure to fully encapsulate the many dynamics that make up the human gait; rather multiple methods, measured through a variety of capture devices, can be used to perform gait recognition. The choice of sensor was highlighted as being particularly important to the design of such a system, and the various methods for achieving gait recognition were expanded up on in the rest of the chapter, where it was shown that gait recognition could be broadly categorized into three different approaches: the MV approach, the WS approach, and the FS approach. Significant research was shown to have gone into each of the three different approaches, in some cases with promising results. Finally, this chapter dealt with some of the key security and privacy concerns that would need to be considered before deploying any gait biometric system to a practical setting.

Looking back on the research relating to the three different gait recognition approaches, one thing that stands out is the relative strength of the WS and FS approaches in comparison with the MV performance. This was particularly true for the approaches that measured footsteps, with nearly every study achieving an error rate of less than 10 %. It is possible that these results are just a reflection of the MV databases being larger and more similar to a real-world environment, with the others being restricted to a laboratory environment; however, it might also be the case that the sensors in contact with the body during a detected gait motion provide more reliability than those capturing a video from a distance. It should be noted that both sensor approaches, when configured to measure the forces of a footstep, produce a signal representation referred as the *Ground Reaction Force* (GRF). In the next chapter, we explore this measure in greater detail and discuss some of the machine learning strategies that have previously been utilized to perform GRF-based gait recognition.

References

1. Addlesee, Michael D., Alan Jones, Finnbar Livesey, and Ferdinando Samaria. 1997. The ORL active floor [sensor system]. *IEEE Personal Communications* 4(5): 35–41.
2. Akaike, Hirotugu. 1983. Information measures and model selection. *Bulletin of the International Statistical Institute* 50: 277–290.
3. Bamberg, Stacy J. Morris, Ari Y. Benbasat, Donna Moxley Scarborough, David E. Krebs, and Joseph A. Paradiso. 2008. Gait analysis using a shoe integrated wireless sensor system. *IEEE Transactions on Information Technology in Biomedicine* 12 (4): 413–423.
4. Bauer, Eric, and Ron Kohavi. 1999. An empirical comparison of voting classification algorithms: bagging, boosting, and variants. *Machine Learning* 36(1–2): 105–139.
5. Bernardo, José M., and Adrian F.M. Smith. 2008. *Bayesian theory*. New York, USA: Wiley.
6. Bishop, Christopher M. 2006. *Pattern recognition and machine learning*, 1st ed. New York, USA: Springer.
7. Bouchrika, Imed, and Mark S. Nixon. 2008. Exploratory factor analysis of gait recognition. *FG '08. 8th IEEE international conference on automatic face & gesture recognition*, Amsterdam, 1–6.
8. Bouchrika, Imed, and Mark S. Nixon. 2007. Model-based feature extraction for gait analysis and recognition. In *Mirage: Computer vision / computer graphics collaboration techniques and applications*, 150–160. Rocquencourt: Inria.
9. Boulgouris, Nikolaos V., Dimitrios Hatzinakos, and Konstantinos N. Plataniotis. 2005. Gait recognition: A challenging signal processing technology for biometric identification. *IEEE Signal Processing Magazine* 22(6): 78–90.
10. Boulgouris, Nikolaos V., Konstantinos N. Plataniotis, and Dimitrios Hatzinakos. 2006. Gait recognition using linear time normalization. *Pattern Recognition* 39(5): 969–979.
11. Brownlee, Jason. 2013. Taxonomy of machine learning techniques. [Online]. http://machinelearningmastery.com/a-tour-of-machine-learning-algorithms/.
12. Calder, Alan, and Steve G. Watkins. 2010. *Information security risk management for ISO27001/ISO27002*. United Kingdom: IT Governance Publishing.
13. Cattin, Philippe C. 2002. Biometric authentication system using human gait. Ph.D. thesis, Swiss Federal Institute of Technology, Zurich, Switzerland.
14. Cavoukian, Ann. 1999. Consumer biometric applications: A discussion paper. Information and Privacy Commissioner, Ontario, Toronto.
15. Chapman, Arthur D. 2005. *Principles and methods of data cleaning—primary species and species-occurrence data*. Copenhagen: Global Biodiversity Information Facility.
16. Cheng, Ming Hsu, Meng Fen Ho, and Chung Lin Huan. 2008. Gait analysis for human identification through manifold learning and HMM. *Pattern Recognition* 41(8): 2541–2553.
17. Dash, Manoranjan, and Huan Liu. 1997. Feature selection for classification. *Intelligent Data Analysis* 1(3): 131–156.
18. Derawi, Mohammad Omar, Patrick Bours, and Kjetil Holien. 2010. Improved cycle detection for accelerometer based gait authentication. In *Sixth international conference on intelligent information hiding and multimedia signal processing*, Darmstadt, 312–317.
19. Derawi, Mohammad Omar, Davrondzhon Gafurov, and Patrick Bours. 2012. Towards continuous authentication based on gait using wearable motion recording sensors. In *Continuous authentication using biometrics: Data, models, and metrics*, ed. Issa Traoré and Ahmed Awad E. Ahmed, Chap. 8, 170–192. IGI Global.

20. Fitzgerald. Michael Nowlan. 2009. Human identification via gait recognition using accelerometer gyro force. http://www.cs.yale.edu/homes/mfn3/pub/mfn_gait_id.pdf.
21. Gafurov, Davrondzhon. 2007. A survey of biometric gait recognition: Approaches, security and challenges. In *Annual Norwegian computer science conference*, Bergen, 19–31.
22. Gafurov, Davrondzhon, Kirsi Helkala, and Søndrol Torkjel. 2006. Biometric gait authentication using accelerometer sensor. *Journal of Computers* 1(7): 51–58.
23. Gafurov, Davrondzhon, and Einar Snekkenes. 2009. Gait recognition using wearable motion recording sensors. *EURASIP Journal on Advances in Signal Processing* 2009(1): 1–16.
24. Goudelis, Georgios, Anastasios Tefas, and Ioanis Pitas. 2010. Intelligent multimedia analysis for emerging biometrics, Chap. 5. In *Intelligent multimedia analysis for security applications*, 97–125. Berlin Heidelberg: Springer.
25. Guyon, Isabelle, Amir Saffari, Dror Gideon, and Gavin Cawley. 2010. Model selection: Beyond the bayesian/frequentist divide. *Journal of Machine Learning Research* 11: 61–87.
26. Huang, Bufu, Meng Chen, Panfeng Huang, and Yangsheng Xu. 2007. Gait modeling for human identification. In *IEEE international conference on robotics and automation*, Roma, 4833–4838.
27. Huang, Bufu, Meng Chen, Weizhong Ye, and Yangsheng Xu. 2006. Intelligent shoes for human identification. In *IEEE international conference on robotics and biomimetics*, Kunming, 601–606.
28. Itai, Akitoshi, and Hiroshi Yasukawa. 2006. Personal identification using footstep based on wavelets. In *International symposium on intelligent signal processing and communication systems*, Totoori, 383–386.
29. Iwama, Haruyuki, Mayu Okumura, Yasushi Makihara, and Yasushi Yagi. 2012. The OU-ISIR gait database comprising the large population dataset and performance evaluation of gait recognition. *IEEE Transaction on Information Forensics and Security* 7(5), 1511–1521.
30. Japkowicz, Nathalie, and Mohak Shah. 2011. *Evaluating learning algorithms: A classification perspective*. New York, NY, USA: Cambridge University Press.
31. Kadane, Joseph B., and Nicole A. Lazar. 2004. Methods and criteria for model selection. *Journal of the American Statistical Association* 99(465): 279–290.
32. Liu, Zongyi, and Sudeep Sarkar. 2006. Improved gait recognition by gait dynamics normalization. *IEEE Transactions on Pattern Analysis and Machine Intelligence* 28(6): 863–876.
33. Lu, Jiwen, and Erhu Zhang. 2007. Gait recognition for human identification based on ICA and fuzzy SVM through multiple views fusion. *Pattern Recognition Letters* 28(16): 2401–2411.
34. Manyika, James et al. 2011. Bigdata: The next frontier for innovation, competition, and productivity. Technical Report, McKinsey Global Institute.
35. Mitchell, Tom. 1997. *Machine learning*. New York, USA: McGraw Hill.
36. Model Selection. 2015. http://bactra.org/notebooks/model-selection.html.
37. Moustakidis, Serafeim P., John B. Theocharis, and Giannis Giakas. 2008. Subject recognition based on ground reaction force measurements of gait signals. *IEEE Transactions on Systems, Man, and Cybernetics-Part B: Cybernetics* 38(6): 1476–1485.
38. Narendra, Kumpati S. and Mandayam A.L. Thathachar. 1989. *Learning automata: An introduction*. Upper Saddle River, NJ, USA: Prentice-Hall, Inc.
39. Nixon, Mark. 2008. Gait biometrics. *Biometric Technology Today* 16(7–8): 8–9.
40. Nixon, Mark S., and John N. Carter. 2006. Automatic recognition by gait. *Preedings of the IEEE* 94(11): 2013–2024.
41. Orr, Robert J., and Gregory D. Abowd. 2000. The smart floor: A mechanism for natural user identification and tracking. In *CHI '00 conference on human factors in computer systems*, The Hague, 275–276.
42. Ratha, Nalini K., Jonathan H. Connell, and Ruud M. Bolle. 2001. Enhancing security and privacy in biometrics-based authentication systems. *IBM Systems Journal* 40(3): 614–634.
43. Rodríguez, Rubén Vera, Nicholas W.D. Evans, Richard P. Lewis, Benoit Fauve, and John S. D. Mason. 2007. An experimental study on the feasibility of footsteps as a biometric. In *15th European signal processing conference (EUSIPCO 2007)*, Poznan, 748–752.

44. Rodríguez, Rubén Vera, John S.D. Mason, and Nicholas W.D. Evans. 2008. Footstep recognition for a smart home environment. *International Journal of Smart Home* 2(2): 95–110.
45. Salarian, Arash et al. 2004. Gait assessment in Parkinson's disease: Toward an ambulatory system for long-term monitoring. *IEEE Transactions on Biomedical Engineering* 51(8): 1434–1443.
46. Samudin, Norshuhada, Wan Noorshahida Mohd Isa, Tomás H. Maul, and Weng Kin Lai. 2009. Analysis of gait features between loaded and normal gait. In *Fifth international conference on signal image technology and internet based systems*, Marrakech, 172–179.
47. Sazonov, Edward S., Timothy Bumpus, Stacey Zeigler, and Samantha Marocco. 2005. Classification of plantar pressure and heel acceleration patterns using neuraln networks. In *IEEE international joint conference on neural networks*, vol. 5, Montreal, 3007–3010.
48. Sewell, Martin. 2007. Feature selection. http://machine-learning.martinsewell.com/feature-selection/.
49. Spranger, Sabastijan, and Damjan Zazula. 2009. Gait identification using cumulants of accelerometer data. In *2nd WSEAS international conference on sensors, and signals and visualization, imaging and simulation and materials science*, Stevens Point, 94–99.
50. Suutala, Jaakko, Kaori Fujinami, and Juha Röning. 2008. Gaussian process person identifier based on simple floor sensors. In *Smart sensing and context third European conference, EuroSSC*, Zurich, 58–68.
51. Suutala, Jaakko, and Juha Röning. 2008. Methods for person identification on a pressure-sensitive floor: Experiments with multiple classifiers and reject option. *Information Fusion Journal, Special Issue on Applications of Ensemble Methods* 9(1): 21–40.
52. Swansea Footstep Recognition Dataset. [Online]. http://eeswan.swan.ac.uk.
53. Tang, Jiexiong, Chenwei Deng, and Guang-Bin Huang. May 2015. Extreme Learning Machine for Multilayer Perceptron. *IEEE Transactions on Neural Networks and Learning Systems* PP(99): 1. DOI:10.1109/TNNLS.2015.2424995.
54. Venkat, Ibrahim, and Philippe De Wilde. 2011. Robust gait recognition by learning and exploiting sub-gait characteristics. *International Journal of Computer Vision* 91(1): 7–23.
55. Wang, Liang, Tieniu Tan, Huazhong Ning, and Weiming Hu. 2003. Silhouette analysis-based gait recognition for human identification. *IEEE Transactions on Pattern Analysis and Machine Intelligence* 25(12): 1505–1518.
56. Witten, Ian H., Eibe Frank, and Mark A. Hall. 2011. *Data mining practical machine learning tools and techniques*, 3rd ed. San Mateo, CA, USA: Morgan Kaufmann.
57. Woodward, John D. et al. 2001. Biometrics: A technical primer, Chap. 2. In *Army biometric applications: Identifying and addressing sociocultural concerns*, Rand, 67–86.
58. Ye, Hong, Syoji Kobashi, Yutaka Hata, Kazuhiko Taniguchi, and Kazunari Asari. 2009. Biometric system by foot pressure change based on neural network. In *39th international symposium on multiple-valued logic*, Naha, 18–23.
59. Yun, Jaeseok. 2011. User identification using gait patterns on UbiFloorII. *Sensors* 11: 2611–2639.
60. Zhou, Xiaoli, and Bir Bhanu. 2006. Feature fusion of face and gait for human recognition at a distance in video. In *Proceedings of IEEE international conference pattern recognition*, Hong Kong, 529–532.

Chapter 3
Gait Biometric Recognition Using the Footstep Ground Reaction Force

The design of a gait-based biometric system is highly dependent on the aspect of gait being studied and may be limited by the tools available for study. The research presented in the previous chapter has suggested that the *Ground Reaction Force*, measured using either a WS or FS approach, may lead to better results than might be achieved using a more conventional MV approach. Consequently, for the purpose of the analysis provided in this book, we have chosen to study GRF-based gait recognition, and for the remainder of the book, we will be performing the analysis on the GRF data acquired via a *floor-mounted force plate* (an FS approach). The GRF is typically represented as one or more discrete-time signals measuring the force exerted by the ground back on the foot at varying points in the footstep. This signal representation of the GRF is advantageous because it opens the gait biometric for study using well-established and leading edge analysis techniques, which have often previously catered to the examination of similar problems in alternate domains. In this chapter, we begin by exploring the GRF in greater detail and proceed to describe the varying analysis techniques used in previous GRF recognition studies. In doing so, we identify several key research gaps, which will be addressed via novel research presented in the chapters that follow.

3.1 The Ground Reaction Force

The GRF provides us with a representation of the forces exerted throughout the duration of a footstep and is typically visualized as one or more plotted time series. Measurements of the GRF are typically collected at even intervals obtained using either the FS approach with *force plates*, or, less commonly, using a *shoe-based* WS approach. At the moment most research has dealt with GRF captured via force plate sensors, but there are projects, including the work of Plantiga Technologies Inc. [10] that are examining the incorporation of GRF recognition into a shoe-based wearable sensor. The variety of technologies available for GRF capture leads to a variety of available data formats; the data available for study may include either an

© Springer International Publishing Switzerland 2016
J.E. Mason et al., *Machine Learning Techniques for Gait Biometric Recognition*,
DOI 10.1007/978-3-319-29088-1_3

overall representation of total aggregate footstep force, or be divided into smaller subcomponents of force that better describe the nuances of the step. The most common representation of the GRF places it within a 3D Cartesian space with dimensions having the force components running along each axis.

A popular representation of the aforementioned GRF component breakdown is the Kistler force plate coordinate system, demonstrated in Fig. 3.1. In this system, the GRF is represented by a three component force vector, with each component reflecting a different aspect of the footstep either perpendicular or parallel to the ground. The *vertical* force component of the footstep shown as **Fz** in Fig. 3.1 represents the vertical acceleration of the body, and is larger when the body is accelerating upward and smaller when the body accelerates downward. The time series vertical GRF (**Fz**) of a single footstep is shown in Fig. 3.2; it has two distinct peaks that correspond first to the phase in the step where the foot impacts the ground, and then to the phase where the foot pushes up off the ground. The

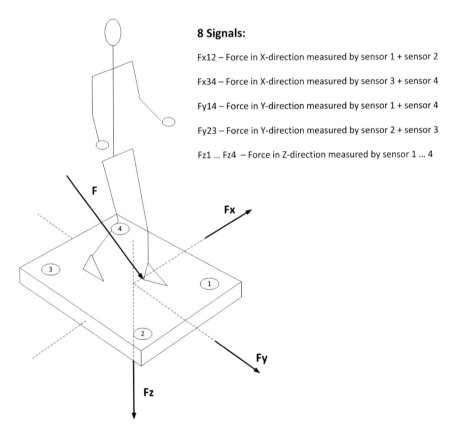

8 Signals:

Fx12 – Force in X-direction measured by sensor 1 + sensor 2

Fx34 – Force in X-direction measured by sensor 3 + sensor 4

Fy14 – Force in Y-direction measured by sensor 1 + sensor 4

Fy23 – Force in Y-direction measured by sensor 2 + sensor 3

Fz1 … Fz4 – Force in Z-direction measured by sensor 1 … 4

Fig. 3.1 Kistler Force Plate Coordinate System [5]. This figure represents the Kistler coordinate system, with the force labeled 'F' representing the stepping force vector. The force plate translates this into its vertical (Fz), anterior–posterior (Fy), and medial–lateral (Fx) components using four different sensors

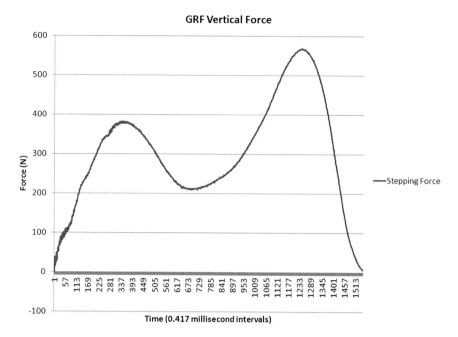

Fig. 3.2 Footstep GRF Vertical Force. This figure demonstrates the GRF vertical force for a footstep

anterior–posterior force, shown as **Fy** in Fig. 3.1 represents the horizontal friction between the foot and the ground. This component, shown in the time series in Fig. 3.3, is largely responsible for horizontal motion, and its peaks and troughs correspond to forward acceleration and impact breaking, respectively. Finally, the *medial–lateral* component of the footstep shown as **Fx** in Fig. 3.1 represents the friction forces perpendicularly to the direction of motion; these forces, shown in the time series in Fig. 3.4 reflect the rotation of the ankle during a footstep. Most researchers have tended to focus on the vertical component of the GRF for recognition; however, studies have indicated that the anterior–posterior and medial–lateral forces also contain valuable subject specific information [4].

Since the footstep GRF was first proposed as a biometric for gait-based recognition in 1997 [12], only a small number of researchers have examined it for its stand-alone recognition capability. The results of some of the key GRF-based gait recognition studies are shown in Table 3.1. Most of these results were discussed in Sect. 2.2 of the previous chapter, but this table provides some greater details with respect to the dataset characteristics and processing techniques used by each. Through the comparison of results from previous GRF-based gait recognition studies with respect to their chosen machine learning techniques, we can get a rough idea of the strength that might be achievable when applying each of their chosen machine learning techniques. This in turn provides a foundation for techniques we have chosen to cover in our own analysis. However, one caveat in

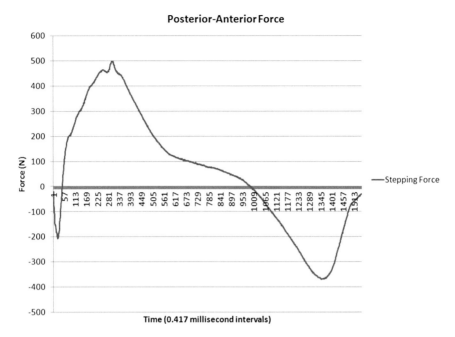

Fig. 3.3 Footstep GRF Posterior–Anterior Force. This figure demonstrates the GRF posterior–anterior force for a footstep

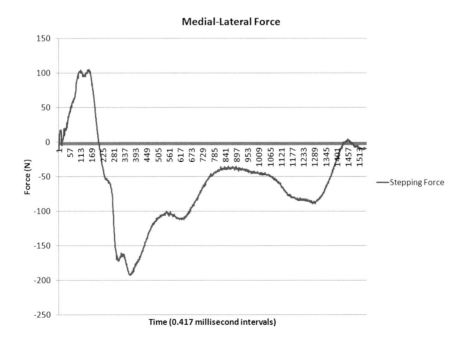

Fig. 3.4 Footstep GRF Medial–Lateral Force. This figure demonstrates the GRF medial–lateral force for a footstep

Table 3.1 GRF Recognition-Related Research

Group/year	Steps/persons	Classifier	Results (%)	Multi-Shoes	Normalized	Samples/person
Addlesee et al. [1]	300/15	HMM	ID rate: 91	No	No	10 steps
Orr and Abowd [9]	1,680/15	NN	ID rate: 93	Yes	Yes	10 steps
Cattin [4]	480/16	Euclid. Distance	EER: 9.4	Yes	No	6 step cycles
Suutala and Röning [13]	440/11	SVM	ID rate: 94	No	No	27 steps
Moustakidis et al. [7]	2,800/40	GK-SVM	ID rate: 98.3	No	Yes	7 steps
Rodriguez et al. [11]	3,174/41	SVM	EER: 9.5	Yes	Yes	40 steps
Mostayed et al. [6]	18/6	Histogram Similarity	EER: 3.3 – 16	No	No	1 step
Rodriguez et al. [12]	3,550/55	SVM	EER: 3	Yes	Yes	40 steps

This table shows the comparison of different approaches to GRF footstep recognition. Results refer to those obtained using a development dataset. *Step cycles* refer to the combination of the right and left footsteps that make up a walking cycle comparing different studies is that they were not all performed over the same dataset, and factors like the size and quality of the dataset can have a significant impact on the performance of the system. It must also be noted that the results demonstrated often come only from the dataset used for development, and, when evaluation datasets are used, performance tends to decrease. This decrease in performance was demonstrated by Rodríguez et al. in [11, 12], with an increase in verification EER of 21 and 330 %, respectively, when evaluation datasets were used instead of development sets. Furthermore, while some studies measured the performance as the ability of a classifier to identify a person using footstep profiles, others were based on the ability of a system to verify a person's identity given its credentials and a footstep. Moreover, in some studies, a large number of footsteps were used in classifier training, which will be shown later in this book to have a significant impact on the recognition performance.

Despite the difficultly in performing direct comparisons, the results shown in Table 3.1 do appear to indicate that the GRF is capable of producing similar, or perhaps better, recognition performance than achieved using the MV-based recognition approach; possibly a reflection of the metric being less prone to obscuring variability (or *covariate factors*). Yet, while GRF features appear to be less susceptible to covariate factors than gait features captured by video [4], there still is a potential for factors like varying *shoe type* or *stepping speed* to reduce the GRF recognition accuracy. Of the eight studies examined, only four attempted recognition using datasets that included multiple shoe types for a single person. Likewise, normalization, a natural technique to reduce the impact of variance such as disparities in measured stepping speed, was also only applied in four studies. The sections that follow expand upon the discussion of how GRF research to date has addressed the *primary emphasis* of this book, the incorporation of machine learning techniques into GRF-based gait analysis for optimal human recognition, as well as the important *secondary emphases*: the impacts of shoe type variance and stepping speed normalization on the recognition performance.

3.2 Feature Extraction

In Sect. 2.1 of the previous chapter, we introduced the concept of a biometric system to perform gait-based human recognition. The first step in building a biometric recognition system involves identifying the most discriminative features that can be extracted from raw data. Ideally, features used for recognition should appear *consistently* for all persons tested, yet show enough *variance* such that there is no overlap in feature space between two or more persons. In reality, finding such features can be a difficult task, particularly for behavioral biometrics, and there is usually at least some degree of *overlap* in feature spaces. In the studies referenced in Table 3.1, four different types of features are presented: geometric features [7, 9, 11–13], holistic features [1, 6, 11, 12], spectral features [4, 13], and wavelet transform features [7].

Geometric features refer to a feature set that is determined using well-recognized geometric attributes like max and min points, as well as statistical attributes like mean and standard deviation. Most studies related to the GRF recognition have used geometric features as either a *primary feature set* for classification, or as a *comparison feature set* to assess the effectiveness of an alternative feature extraction technique. In [9], a biometric recognition system was built using the mean, standard deviation, area under the curve, and extrema (min/max) points for a footstep GRF graph; it was noted in the study that the mean and standard deviation appeared to show the highest discriminative power. An attempt to *optimize* the geometric feature set was made in [12]. In this study, the geometric feature set for the footstep GRF graph contained extrema points, the distances between extrema points, the area under the curve, the norm, the mean, the length, and the standard deviation. To determine the best features an exhaustive search was performed, searching for the combination of features that minimized the verification EER across a development dataset. The result was a reduction in feature dimensionality from 42 features to 17. The optimal features included 11 extrema point features, 2 area features, 2 norm features, and 2 standard deviation features, for which a 27 % increase in performance was noted. While the optimized geometric features showed a significant improvement in performance, papers [7, 11, 12] demonstrated comparatively better results with different feature extraction approaches, and in [13], it was found that combining the geometric feature set with a set of another feature type produced a significant increase in performance.

Holistic features have appeared as a promising alternative to geometric features in several GRF recognition studies. The holistic approach to feature extraction generally involves dividing the raw data into some arbitrary number of equally spaced points (or samples) then allowing the most discriminative points to reveal themselves to a classifier. The simplest technique, used in [1], involves passing the data directly into a classifier. This approach can run into problems, as it tends to lead to an incredibly large feature space dimensionality, but there are solutions. In [6], the dataset was simplified using a representative histogram. Yet, the most effective simplification technique was demonstrated in [11], where the raw holistic feature set originally contained 4,200 features but was simplified using the *Principal Component Analysis* approach. The use of PCA made it possible to generate a *reduction in dimensionality* while retaining as much *information* about the original data as possible. In applying PCA, the study demonstrated that the first 80 principal components contained 96 % of the original information, while reducing dimensionality by 98 %. When performance of these 80 holistic features was compared against the performance of a geometric feature set, a 46 % increase in performance was observed.

Another alternative to the geometric feature set was proposed in [7]. This approach performed feature extraction by first translating the GRF data to the *time–frequency domain* using a wavelet packet transform, then extracting the 100 most discriminative features from the data using a form of optimized *Wavelet Packet Decomposition*. Converting the data to the time–frequency domain allowed the complex information and patterns associated with the GRF to be represented in a simpler form. This characteristic proved very useful for classifying footstep GRF samples with *walking speeds that differed* from those the system was trained on; in these instances, the increase in recognition performance, when compared with the recognition results from a 16 feature standard geometric set, was as high as 66 %. This compared similarly with the performance increase observed over the geometric feature set when training and testing data came from the *same walking speed* range, which was also around 66 %.

Additionally, the *special features*, as well as the features extracted from the *frequency domain* have proven useful in two previous GRF-based gait recognition studies. In [4], a feature set, derived from the windowed Power Spectral Density (PSD) function of the derivative GRF was suggested to provide stronger recognition ability than could be achieved with a geometric feature set. The PSD function is particularly useful because it shows the strength of *energy* variations as a function of frequency. By identifying the frequencies at which the variations are strong, this function could make it easier for a classifier to identify the most important features. To capture a low dimensionality feature set from the PSD function, a novel Generalized PCA (GPCA) technique was used, and it was found that the best 10 dimensions (features) contained more than 90 % of the dataset variance. The technique produced a reasonably strong verification EER of 9.4 %, however, no equivalent geometric feature set was tested on the dataset, so it was not possible to make a direct comparison between the two techniques. In [13], two holistic feature sets, derived from the frequency domain of the GRF and its derivative function, were combined with a geometric feature set. While, on their own, these spectral feature sets performed worse than the geometric set in this study, when all three sets were combined the resulting performance was 36 % better than the best stand-alone result.

Not only did the studies in Table 3.1 vary in their approaches to feature extraction, but they also varied in the components and *quality* of data collected. In [9, 11–13], an averaged GRF was obtained and examined, rather than one split into its three vector components. In [1, 4, 6], only the vertical component of the GRF was used for final classification analysis; both [4, 6], regarded it as the best discriminator. The only study that used all three components of the GRF for classification was [7]. No study examined GRF classification for more information rich data samples involving the output from four or more GRF component *output signals*, opening the possibility for further study into GRF feature extraction using higher quality data. Furthermore, since previous studies did not share the same dataset, a better relative comparison of feature set performance could be achieved by comparing the feature extraction techniques of different studies on the same dataset.

3.3 Normalization

It is very difficult for a person to perform the same action twice with no measureable difference between the two attempts. When collecting a feature set for the footstep GRF, differences in walking speed can have a large impact on the timing and amplitude of the extracted features. Unfortunately walking speed, and by extension *stepping speed* appear to be extremely difficult to regulate with high precision, even in a controlled experiment. One technique that can be used to account for the natural variation in data is *normalization*. Normalization assumes that some sort of relationship exists between two or more variables, and, by using this relationship, variables can be projected onto the same reference point for a more accurate comparison.

Of the studies referenced in Table 3.1, only four applied normalization to their dataset; two of these were from the same research group [11, 12]. None of the studies covered their chosen normalization techniques in detail, and it appeared that only simple normalization techniques were used. In [9], data normalization was mentioned, but no detail was given regarding the technique used or the target of the normalization. In [7], the data was normalized around the weight of test subjects so, when loads of 5 and 10 % of the subjects body weight were added during testing, it was possible to adjust the feature set to a common weight reference point. The study also used a simple resampling-based *Linear Time Normalization* (LTN) technique to address differences in step sample length (duration); however, the feature extraction technique also focused on capturing features less sensitive to walking speed variation and no non-normalized results were presented for reference. Finally, in [11, 12], feature sets were normalized with respect to the *absolute maximum value* of the GRF footstep profile; this simple approach would appear to account for variations in stepping force but not step duration.

While no study dealing directly with GRF for recognition examined the actual effects of using normalization to address differences in walking speed, there is evidence from other related studies suggesting that such an approach may achieve better recognition results. A study examining gait using the MV approach revealed that applying LTN to the feature data improves the identification performance over non-normalized feature sets by 8–20 % [3]; this result implied the existence of an identifiable relationship between *walking speed* and observable *gait characteristics*. The impact of walking speed on GRF was also examined in [14] as part of a human kinetics study that analyzed its relationship with the vertical GRF component. The study examined the difference in the amplitude of the vertical GRF across three different walking speeds for 20 volunteers. It was found that: the *maximum amplitude* increased by 2 % when walking at a normal speed compared to a slow speed, it increased by 6 % when walking at a fast speed compared to a normal speed, and by 9 % when walking at a fast speed compared to a slow speed. The identification of such a clear relationship between walking speed and GRF supports the need for further investigation into utilizing this relationship to improve the recognition results.

3.4 Classification Approaches

In biometric recognition, *classifiers* are the algorithms that take a feature set as input then attempt to either *assign* it an identity, or *verify* that it corresponds to a provided identity. Classifiers can be categorized according to two different models:

generative models and discriminative models [2]. *Generative* classifiers involve estimating an input distribution, then modeling the class conditional densities, and finally calculating the *posterior class probability* via *Bayes' rule* (this being the probability that a set of features corresponds to a given class); for instance, to learn the posterior class probability function $P(X|Y)$, where X is a class and Y is a feature, a generative classifier would be used to estimate the *a priori* probability for each class $P(X)$ and class conditional probability $P(Y|X)$, then Bayes' rule will be applied to get the intended result. Conversely, *discriminative* classifiers are based on decision boundaries that *minimize the classification error loss* over the true class conditional probabilities and model posterior class probabilities directly or learn a direct mapping to class labels [8]; using the previous example, a discriminative classifier might attempt to determine $P(X|Y)$ directly. Of these two approaches, discriminative classifiers have generally proven to yield a better performance for footstep recognition [13]. In the studies presented in Table 3.1, nine different classifiers were tested and the most successful methods were identified in the classifier column. In these studies, only three generative classifiers (Maximum Likelihood (ML), LDA, HMM) were studied, while the remaining six were discriminative (KNN, SVM, MLP, Radial Basis Function (RBF), Learning Vector Quantization (LVQ), C4.5). In most studies, a single classifier was trained to make decisions across the full feature space. However, in [13], three different instances of a chosen classifier were trained using three distinctive regions in the feature space, and the posterior probabilities returned by the three classifiers were fused into a single probability using *combination rules*; the result was a 46 % decrease in error by the strongest classifier.

The most commonly used classifiers in the studies of Table 3.1 were variants of the *K-Nearest Neighbor* classifier. The KNN classifier is a simple algorithm that assigns a feature set to the closest known identity (class), measured as the *distance* between a known feature set and the input feature set being classified. Variants of this classifier include the histogram similarity approach described in [6], and the Euclidean distance approach described in [4]. In [9], a relatively high identification rate of 93 % was achieved using simple KNN, but, in [7, 11, 13], other classifiers were compared with KNN and showed better recognition performance.

After KNN, the next most widely used, and most successful GRF classifier in Table 3.1, was the *Support Vector Machine* classifier. The SVM classifier is a supervised learning method that constructs a *hyperplane* or set of hyperplanes in a high dimensional space, making the separation of complex classes easier. In [7, 11, 13] this classifier generally demonstrated the strongest performance when compared against a number of other classifiers, with a performance increase ranging from 6 to 60 % over the standard KNN classifier. However, in [7], the *Linear Discriminant Analysis*, a classification technique that searches for the linear combination of features to best separate two or more classes, demonstrated similar performance to the SVM classifier when large feature sets were tested. Also, in [13], a *Multilayer Perceptron* classifier demonstrated only slightly weaker identification rates than that of the SVM classifier. None of the remaining classifiers covered by [7, 13] (RBF [7, 13], LVQ [13], ML [7], C4.5 [7]) performed much better than the KNN classifier, while the HMM classifier, studied in [1], has not appeared in more recent GRF recognition research.

Clearly, the choice of classifier plays a strong role in the GRF-based gait recognition performance, but classifiers must be trained and the *number of samples* used for training can also affect the performance. In the studies mentioned in Table 3.1, the number of samples used for training ranged from 1 [6] to 40 [11, 12] GRF samples per person. However, only the study in attempted to find an optimal number of footsteps for classifier training. In this study, recognition was tested across 1–63 training steps and performance was demonstrated to increase substantially until about the 40th step, after which it leveled off. While 40 training steps appeared optimal for this particular study, it is important to note that since each study used a different dataset, the optimal number of training samples for one study cannot be expected to be equivalent in another.

The *number of footsteps* used per single classification attempt is another factor that can affect the recognition performance. Only two of the studies in Table 3.1 examined *multi-footstep* classification. In [4], training and classification were done using two-step cycles (the right and left steps that form a walking cycle). In [13], multi-footstep classification was compared directly with single footstep classification and a 76 % increase in performance over single-step classification was observed using two-step classification, while a 95 % increase in performance was observed with four-step classification. The study also applied a *sample-rejection* strategy to ignore the unreliable data samples from training and testing. Then, using the three-footstep classification with the most unreliable 1 % of the dataset rejected, the study achieved a 100 % identification rate.

One final classification consideration results in best practices when demonstrating the classification results. In [11, 12], the separation of test data into a *development* and *evaluation* set was emphasized. When building a classifier, the development dataset is used to optimize the algorithm to the chosen feature set, while the evaluation set contains previously unseen data, and is used to confirm the results of the development set. Many of the studies demonstrated in Table 3.1 did not use an evaluation dataset, so, for the purpose of making better comparisons, all results demonstrated in the table referred to those obtained using a development dataset. Furthermore, because footstep GRF-based gait recognition is such a new field of study, most research has been restricted to relatively small datasets compared to more traditional biometrics.

Classification algorithms fall into a broader field of machine learning, and have received extensive research over the past few decades. Recognition using the GRF has clearly benefited from the development of classifiers in related biometric research, and, it is apparent from the studies in Table 3.1 that most of the strongest known classifiers have already been attempted by existing research. However, this area of research is constantly evolving and there is always room for testing previously untested classification algorithms for GRF recognition. Moreover, since most datasets previously used in GRF recognition were built on limited, low resolution sensors, it is possible that some algorithms may show an increase in performance and/or a lower training cost given a more descriptive dataset.

3.5 Shoe Type

Even with a highly discriminative normalized feature set and a strong classification algorithm, there will always be some level of variability in human gait that makes GRF-based gait recognition difficult. One such source of variability can arise from the use of a *different shoe* for classifier training than was used for identification or verification. Unlike stepping speed variance, which can be calculated directly from a GRF step time series, there is no way to determine that an individual is wearing a new shoe type based on the footstep GRF signature alone, therefore normalization by shoe type does not seem possible without further environment information. Of the eight studies in Table 3.1, only four examined a sample dataset containing more than one shoe types.

The effect of variable shoe type on the classification performance is debatable. In [9], multiple shoe types were captured for classification and it was concluded that shoe type has little effect on the ability to perform footstep GRF identification. However, this study never indicated whether the multiple shoe types were used for the same person or simply across the whole test group, nor did it provide any information as to whether a different shoe type was used for training than for testing. Conversely, in [4], a more detailed analysis revealed that testing with shoe types *unknown* to the classifier could potentially have a very negative effect on recognition performance. Poor performance was demonstrated by a Euclidean distance classifier when new shoe types appeared in test data, however, when multiple shoe types were used for both training and testing there was a considerable improvement in performance. Finally, the remaining two studies, [11, 12], both mentioned that two or more shoe types per person were included in their datasets, yet no analysis was done to study effects of these shoe types on classifier performance. Neither study had poor enough results to suggest that using multiple shoe types was having a very negative impact on classifier performance, though not enough information was provided to completely rule it out.

So, while a small group of researchers have studied footstep GRF recognition using datasets with multiple shoe types, only a single study by Cattin [4], went into some detail regarding the effect of shoe type on recognition. Moreover, even in [4], critical pieces of information were missing from analysis. For instance, Cattin found that classification performance was weaker in a multi shoe type dataset than in a single shoe type dataset; however, he did not specify whether the choice of shoe type altered stepping speed (a factor that could potentially be mitigated by normalization). Furthermore, Cattin only examined the effect of shoe type on classification using a Euclidean distance classifier. He discovered that training the classifier with multiple shoe types could increase recognition performance across a multi shoe type dataset, but that may not always be an option in a real world environment. It remains to be seen whether a stronger classification algorithm, different feature set, and stepping speed normalization could potentially mitigate the performance decrease that appears when performing classification on a footstep with a shoe type *unknown* to the classifier.

3.6   The Demonstrative Experiment

In the preceding sections, we examined GRF-based gait recognition and identified techniques that have previously been used to accomplish such recognition, while also highlighting potential shortcomings in previous analyses. To illustrate how GRF-based gait recognition techniques can be put into practice and address some of the previously identified research gaps, the remainder of this book is organized around a *demonstrative experiment*. This demonstrative experiment is divided into two parts. The first part of the experiment, presented in Chaps. 4 through 6 examines each of the individual biometric system components and machine learning techniques in greater detail, and shows how each technique can be *optimized* to best work with the GRF data. For this part of the experiment, we work with a 100 sample GRF development dataset generated via recorded footsteps over the 8-signal Kistler force plate shown in Fig. 3.1 (see Sect. 7.3 of Chap. 7 for more details). The second part of the demonstrative experiment begins in Chap. 7, where the optimized recognition techniques of Chaps. 4 through 6 are organized into an experiment to *evaluate* the various GRF-based biometric system configurations, and continues through Chaps. 8 and 9; this part of the experiment uses a larger 299 sample evaluation dataset to mitigate potential development training bias. In addition to performing a comparison of various recognition techniques, the evaluation part of this experiment seeks to address our objective assertions with respect to the impacts of normalization and shoe type of GRF recognition. A high level summary of the demonstrative experiment breakdown is presented in Fig. 3.5.

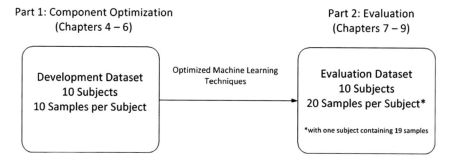

Fig. 3.5 The Demonstrative Experiment. This figure presents a high-level summary of the demonstrative experiment used by this book to illustrate the application of GRF-based gait recognition

3.7 Summary

This chapter presented an overview of the field of GRF-based gait biometric recognition and described how it has been dealt with in previous research. While the first section of this chapter illustrated the properties and terminology that shape the GRF, the remainder of the chapter examined machine learning techniques used in performing GRF recognition and potential areas of concern. In assessing research relevant to GRF recognition, eight different studies were identified and compared according to the set of criteria in Table 3.1. Upon review, the research was found to be lacking with regards to footstep speed normalization and shoe type analysis, while the research areas of feature extraction and classification also left room for new experimentation.

None of the studies listed in this chapter attempted to assess the impact of normalizing the GRF features as a function of stepping speed. And only a single study examined the effect of shoe type on GRF recognition in any detail, leaving a lot of room for further analysis. Feature extraction and classification were more thoroughly covered by existing research, but could still benefit from the use of a more descriptive dataset, a better cross comparison of approaches, and the trial of a previously untested classifier type. To address the identified research gaps and expand upon the work of previous GRF-based gait recognition researchers, this chapter concluded by laying out a demonstrative experiment. This experiment was divided into two parts, with the first exploring the inner workings of a GRF-based biometric system and the second putting the system to the test to determine its best configurations, while simultaneously verifying our assertions regarding normalization and shoe type. The next chapter begins our exploration of the GRF-based gait biometric system, examining what is arguably the most important component when dealing with large complicated data samples like those of the GRF, the *feature extractor*.

References

1. Addlesee, Michael D., Alan Jones, Finnbar Livesey, and Ferdinando Samaria. 1997. The ORL active floor (sensor system). *IEEE Personal Communications* 4(5): 35–41.
2. Bouchard, Guillaume, and Bill Triggs. 2004. The trade-off between generative and discriminative classifiers. In *International Conference on Computational Statistics*, 721–728. Prague.
3. Boulgouris, Nikolaos V., Konstantinos N. Plataniotis, and Dimitrios Hatzinakos. 2006. Gait recognition using linear time normalization. *Pattern Recognition* 39(5): 969–979.
4. Cattin, Philippe C. 2002. Biometric authentication system using human gait, Ph.D. Thesis 2002, Zurich, Switzerland: Swiss Federal Institute of Technology.
5. Kistler force plate formulae. [Online]. http://isbweb.org/software/movanal/vaughan/kistler.pdf.
6. Mostayed, Ahmed, Sikyung Kim, Mohammad Mynuddin Gani Mazumder, and Se Jin Park. 2008. Foot step based person identification using histogram similarity and wavelet decomposition. In *International Conference on Information Security and Assurance*, 307–311, Busan.

7. Moustakidis, Serafeim P., John B. Theocharis, and Giannis Giakas. 2008. Subject recognition based on ground reaction force measurements of gait signals. *IEEE Transactions on Systems, Man, and Cybernetics-Part B: Cybernetics* 38(6): 1476–1485.
8. Ng, Andrew Y., and Michael I. Jordan. 2001. On discriminative vs. generative classifiers: A comparison of logistic regression and naive Bayes. In *Advances in Neural Information Processing Systems 14 (NIPS 2001)*, 841–848. Vancouver.
9. Orr, Robert J, and Gregory D. Abowd. 2000. The smart floor: A mechanism for natural user identification and tracking. In *CHI '00 Conference on Human Factors in Computer Systems*, 275–276, The Hague.
10. Plantiga. [Online]. http://www.plantiga.com.
11. Rodríguez, Rubén Vera, Nicholas W. D. Evans, Richard P. Lewis, Benoit Fauve, and John S. D. Mason. 2007. An experimental study on the feasibility of footsteps as a biometric. In *15th European signal processing conference (EUSIPCO 2007)*, 748–752, Poznan.
12. Rodríguez, Rubén Vera, John S. D. Mason, and Nicholas W. D. Evans. 2008. Footstep recognition for a smart home environment. *International Journal of Smart Home* 2(2): 95–110.
13. Suutala, Jaakko, and Juha Röning. 2008. Methods for person identification on a pressure-sensitive floor: Experiments with multiple classifiers and reject option. *Information Fusion Journal, Special Issue on Applications of Ensemble Methods 9* 9(1): 21–40.
14. Taylor, Amanda J., Hylton B. Menz, and Anne-Maree Keenan. 2004. The influence of walking speed on plantar pressure measurements using the two-step gait initiation protocol. *The Foot* 14(1): 49–55.

Chapter 4
Feature Extraction

Discovering the gait characteristics that have a unique range in values for any given person would make it possible, in the absence of spoofing, to perform recognition with perfect accuracy using even the simplest of classifiers. When dealing with data samples containing thousands of recorded values (also referred to as *dimensions*), leaving the identification of these characteristics to classifiers alone can be computationally expensive and potentially lead to classifier *overfitting*, which occurs when undesirable characteristics, such as noise, are misinterpreted as being significant during training. One way to address these issues is to preprocess the data using a technique known as *feature extraction*.

Feature extraction aims to represent the characteristics that best distinguish the original dataset in a *reduced dimensional space*. The objectives of feature extraction are closely aligned with those of data compression. However, the data compression requirement that enough information be retained to be able to reconstruct the original dataset to some chosen degree of accuracy does not apply to feature extraction. For the purpose of GRF-based gait recognition, the goal is to extract a feature set that minimizes the degree of feature space overlap between any two people. This chapter presents the principles behind four different GRF feature extraction techniques and explains how each of them was configured to accommodate the eight-signal Kistler GRF representation discussed in the previous chapter. Additionally, the work demonstrated in this chapter examines a number of methods used to optimize each extraction technique for a better recognition performance, building upon the work done in previous research.

4.1 Geometric

A number of previous studies [12, 13, 15, 16, 18] have identified *spatial* and *statistical* footstep GRF characteristics that can be extracted to form a feature space. These *heuristically* derived characteristics are referred to as *geometric features*. Spatial features include specific data measurements such as the position of local maxima or displacement between two points of interest, while statistical features reflect the properties of the dataset taken as a whole and include measurements like

© Springer International Publishing Switzerland 2016
J.E. Mason et al., *Machine Learning Techniques for Gait Biometric Recognition*,
DOI 10.1007/978-3-319-29088-1_4

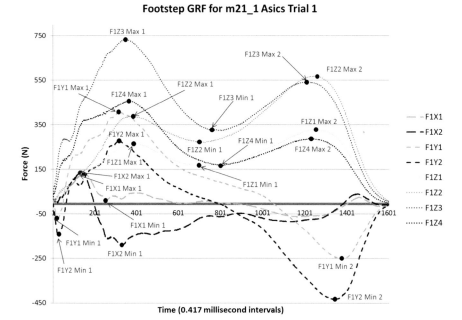

Fig. 4.1 Footstep GRF geometric points of interest. This diagram demonstrates the set of heuristically derived *points of interest* across our eight GRF signals. Each point is identified with a dot and given a descriptive label

the mean GRF value. The sample shown in Fig. 4.1 demonstrates the force values of the 8 output signals over the course of a footstep and has identified the points of interest: the local minima and maxima found to be consistent enough to be used as features.

Using the *points of interest* identified in Fig. 4.1 together with the geometric features proposed in [16], our research was able to identify 538 potential geometric features across the eight output signals. Of these features, 506 were *spatial* characteristics and 32 were *statistical* characteristics. The spatial features included *time* and *force* values for all 22 extrema points identified in Fig. 4.1, the 231 displacements in time between every pair of extrema points, and the 231 displacements in force between every pair of extrema points. The statistical features were restricted to force measurements only and included the mean values, areas under the curve, standard deviations, and norms for each of the 8 output signals. A breakdown of the features is shown in Table 4.1.

Having identified the geometric features of interest, the next challenge was to construct an algorithm capable of *extracting* these features from the GRF curves. The greatest difficulty in this regard involved locating desirable local extrema points on a signal containing a substantial level of noise. To acquire these points, two procedures were developed: a *local maxima locator* and a *local minima locator*. Initially the procedures were set up to accept some initialization point from

Table 4.1 Geometric GRF features

Spatial point features

Output signal	Measure	Features
F1X1	Force	Max1, Min1
F1X1	Time	Max1, Min1
F1X2	Force	Max1, Min1
F1X2	Time	Max1, Min1
F1Y1	Force	Min1, Max1, Min2
F1Y1	Time	Min1, Max1, Min2
F1Y2	Force	Min1, Max1, Min2
F1Y2	Time	Min1, Max1, Min2
F1Z1	Force	Max1, Min1, Max2
F1Z1	Time	Max1, Min1, Max2
F1Z2	Force	Max1, Min1, Max2
F1Z2	Time	Max1, Min1, Max2
F1Z3	Force	Max1, Min1, Max2
F1Z3	Time	Max1, Min1, Max2
F1Z4	Force	Max1, Min1, Max2
F1Z4	Time	Max1, Min1, Max2

Spatial displacement features

Output signal(s)	Features
F1X1	$D\left(\frac{F1X1\ \text{Force}}{2}\right),\ D\left(\frac{F1X1\ \text{Time}}{2}\right)$
F1X1-F1X2	$D\left(\frac{F1X1\ \text{Force} \cup F1X2\ \text{Force}}{2}\right) - \left(\frac{F1X1\ \text{Force}}{2}\right) - \left(\frac{F1X2\ \text{Force}}{2}\right)$
F1X1-F1Y1	$D\left(\frac{F1X1\text{Force} \cup F1Y1\ \boldsymbol{Force}}{2}\right) - \left(\frac{F1X1\ \text{Force}}{2}\right) - \left(\frac{F1\mathbf{Y}1\ \text{Force}}{2}\right),$
F1X1-F1Y2	$D\left(\frac{F1X1\ \boldsymbol{Force} \cup F1Y2\ \text{Force}}{2}\right) - \left(\frac{F1X1\ \text{Force}}{2}\right) - \left(\frac{F1\mathbf{Y}2\ \text{Force}}{2}\right),$
F1X1-F1Z1	$D\left(\frac{F1X1\ \text{Force} \cup F1Z1\ \text{Force}}{2}\right) - \left(\frac{F1X1\ \text{Force}}{2}\right) - \left(\frac{F1\mathbf{Z}1\ \text{Force}}{2}\right),$
F1X1-F1Z2	$D\left(\frac{F1X1\ \text{Force} \cup F1Z2\ \text{Force}}{2}\right) - \left(\frac{F1X1\ \text{Force}}{2}\right) - \left(\frac{F1\mathbf{Z}2\ \text{Force}}{2}\right),$
F1X1-F1Z3	$D\left(\frac{F1X1\ \text{Force} \cup F1Z3\ \text{Force}}{2}\right) - \left(\frac{F1X1\ \text{Force}}{2}\right) - \left(\frac{F1\mathbf{Z}3\ \text{Force}}{2}\right),$
F1X1-F1Z4	$D\left(\frac{F1X1\ \text{Force} \cup F1Z4\ \text{Force}}{2}\right) - \left(\frac{F1X1\ \text{Force}}{2}\right) - \left(\frac{F1\mathbf{Z}4\ \text{Force}}{2}\right),$
F1X2	$D\left(\frac{F1X2\ \text{Force}}{2}\right),\ D\left(\frac{F1X2\ \text{Time}}{2}\right)$

(continued)

Table 4.1 (continued)

Spatial displacement features

Output signal(s)	Features
F1X2-F1Y1	$D\left(\dfrac{F1X2\,\text{Force} \cup \text{F1Y1 Force}}{2}\right) - \left(\dfrac{F1X2\,\text{Force}}{2}\right) - \left(\dfrac{F1\mathbf{Y}1\,\text{Force}}{2}\right)$,
F1X2-F1Y2	$D\left(\dfrac{F1X2\,\text{Force} \cup \text{F1Y2 Force}}{2}\right) - \left(\dfrac{F1X2\,\text{Force}}{2}\right) - \left(\dfrac{F1\mathbf{Y}2\,\text{Force}}{2}\right)$,
F1X2-F1Z1	$D\left(\dfrac{F1X2\,\text{Force} \cup \text{F1Z1 Force}}{2}\right) - \left(\dfrac{F1X2\,\text{Force}}{2}\right) - \left(\dfrac{F1\mathbf{Z}1\,\text{Force}}{2}\right)$,
F1X2-F1Z2	$D\left(\dfrac{F1X2\,\text{Force} \cup \text{F1Z2 Force}}{2}\right) - \left(\dfrac{F1X2\,\text{Force}}{2}\right) - \left(\dfrac{F1\mathbf{Z}2\,\text{Force}}{2}\right)$,
F1X2-F1Z3	$D\left(\dfrac{F1X2\,\text{Force} \cup \text{F1Z3 Force}}{2}\right) - \left(\dfrac{F1X2\,\text{Force}}{2}\right) - \left(\dfrac{F1\mathbf{Z}3\,\text{Force}}{2}\right)$,
F1X2-F1Z4	$D\left(\dfrac{F1X2\,\text{Force} \cup \text{F1Z4 Force}}{2}\right) - \left(\dfrac{F1X2\,\text{Force}}{2}\right) - \left(\dfrac{F1\mathbf{Z}4\,\text{Force}}{2}\right)$,
F1Y1	$D\left(\dfrac{F1\mathbf{Y}1\,\text{Force}}{2}\right), \; D\left(\dfrac{F1\mathbf{Y}1\,\text{Time}}{2}\right)$
F1Y1-F1Y2	$D\left(\dfrac{F1Y1\,\text{Force} \cup \text{F1Y2 Force}}{2}\right) - \left(\dfrac{F1\mathbf{Y}1\,\text{Force}}{2}\right) - \left(\dfrac{F1\mathbf{Y}2\,\text{Force}}{2}\right)$,
F1Y1-F1Z1	$D\left(\dfrac{F1Y1\,\text{Force} \cup \text{F1Z1 Force}}{2}\right) - \left(\dfrac{F1\mathbf{Y}1\,\text{Force}}{2}\right) - \left(\dfrac{F1\mathbf{Z}1\,\text{Force}}{2}\right)$,
F1Y1-F1Z2	$D\left(\dfrac{F1Y1\,\text{Force} \cup \text{F1Z2 Force}}{2}\right) - \left(\dfrac{F1\mathbf{Y}1\,\text{Force}}{2}\right) - \left(\dfrac{F1\mathbf{Z}2\,\text{Force}}{2}\right)$,
F1Y1-F1Z3	$D\left(\dfrac{F1Y1\,\text{Force} \cup \text{F1Z3 Force}}{2}\right) - \left(\dfrac{F1\mathbf{Y}1\,\text{Force}}{2}\right) - \left(\dfrac{F1\mathbf{Z}3\,\text{Force}}{2}\right)$,
F1Y1-F1Z4	$D\left(\dfrac{F1Y1\,\text{Force} \cup \text{F1Z4 Force}}{2}\right) - \left(\dfrac{F1\mathbf{Y}1\,\text{Force}}{2}\right) - \left(\dfrac{F1\mathbf{Z}4\,\text{Force}}{2}\right)$, $D\left(\dfrac{F1Y1\,\text{Time} \cup F1Z4\,\text{Time}}{2}\right) - \left(\dfrac{F1Y1\,\text{Time}}{2}\right) - \left(\dfrac{F1Z4\,\text{Time}2}{2}\right)$
F1Y2	$D\left(\dfrac{F1\mathbf{Y}2\,\text{Force}}{2}\right), \; D\left(\dfrac{F1\mathbf{Y}2\,\text{Time}}{2}\right)$
F1Y2-F1Z1	$D\left(\dfrac{F1Y2\,\text{Force} \cup \text{F1Z1 Force}}{2}\right) - \left(\dfrac{F1\mathbf{Y}2\,\text{Force}}{2}\right) - \left(\dfrac{F1\mathbf{Z}1\,\text{Force}}{2}\right)$, $D\left(\dfrac{F1Y2\,\text{Time} \cup \text{F1Z1 Time}}{2}\right) - \left(\dfrac{F1\mathbf{Y}2\,\text{Time}}{2}\right) - \left(\dfrac{F1Z1\,\text{Time}}{2}\right)$
F1Y2-F1Z2	$D\left(\dfrac{F1Y2\,\text{Force} \cup \text{F1Z2 Force}}{2}\right) - \left(\dfrac{F1\mathbf{Y}2\,\text{Force}}{2}\right) - \left(\dfrac{F1\mathbf{Z}2\,\text{Force}}{2}\right)$, $D\left(\dfrac{F1Y2\,\text{Time} \cup \text{F1Z2 Time}}{2}\right) - \left(\dfrac{F1\mathbf{Y}2\,\text{Time}}{2}\right) - \left(\dfrac{F1Z2\,\text{Time}}{2}\right)$

(continued)

Table 4.1 (continued)

Spatial displacement features

Output signal(s)	Features
F1Y2-F1Z3	$D\left(\dfrac{F1Y2\,\text{Force}\cup F1Z3\,\text{Force}}{2}\right)-\left(\dfrac{F1\mathbf{Y}2\,\text{Force}}{2}\right)-\left(\dfrac{F1\mathbf{Z}3\,\text{Force}}{2}\right),$ $D\left(\dfrac{F1Y2\,\text{Time}\cup F1Z3\,\text{Time}}{2}\right)-\left(\dfrac{F1\mathbf{Y}2\text{Time}}{2}\right)-\left(\dfrac{F1Z3\,\text{Time}}{2}\right)$
F1Y2-F1Z4	$D\left(\dfrac{F1Y2\,\text{Force}\cup F1Z4\,\text{Force}}{2}\right)-\left(\dfrac{F1\mathbf{Y}2\,\text{Force}}{2}\right)-\left(\dfrac{F1\mathbf{Z}4\,\text{Force}}{2}\right),$ $D\left(\dfrac{F1Y2\,\text{Time}\cup F1Z4\,\text{Time}}{2}\right)-\left(\dfrac{F1\mathbf{Y}2\text{Time}}{2}\right)-\left(\dfrac{F1Z4\,\text{Time}}{2}\right)$
F1Z1	$D\left(\dfrac{F1Z1\,\text{Force}}{2}\right),\ D\left(\dfrac{F1Z1\,\text{Time}}{2}\right)$
F1Z1-F1Z2	$D\left(\dfrac{F1Z1\,\text{Force}\cup F1Z2\,\text{Force}}{2}\right)-\left(\dfrac{F1\mathbf{Z}1\,\text{Force}}{2}\right)-\left(\dfrac{F1Z2\,\text{Force}}{2}\right),$ $D\left(\dfrac{F1Z1\,\text{Time}\cup F1Z2\,\text{Time}}{2}\right)-\left(\dfrac{F1\mathbf{Z}1\text{Time}}{2}\right)-\left(\dfrac{F1Z2\,\text{Time}}{2}\right)$
F1Z1-F1Z3	$D\left(\dfrac{F1Z1\,\text{Force}\cup F1Z3\,\text{Force}}{2}\right)-\left(\dfrac{F1\mathbf{Z}1\,\text{Force}}{2}\right)-\left(\dfrac{F1Z3\,\text{Force}}{2}\right),$ $D\left(\dfrac{F1Z1\,\text{Time}\cup F1Z3\,\text{Time}}{2}\right)-\left(\dfrac{F1\mathbf{Z}1\text{Time}}{2}\right)-\left(\dfrac{F1Z3\,\text{Time}}{2}\right)$
F1Z1-F1Z4	$D\left(\dfrac{F1Z1\,\text{Force}\cup F1Z4\,\text{Force}}{2}\right)-\left(\dfrac{F1\mathbf{Z}1\,\text{Force}}{2}\right)-\left(\dfrac{F1Z4\,\text{Force}}{2}\right),$ $D\left(\dfrac{F1Z1\,\text{Time}\cup F1Z4\,\text{Time}}{2}\right)-\left(\dfrac{F1\mathbf{Z}1\text{Time}}{2}\right)-\left(\dfrac{F1Z4\,\text{Time}}{2}\right)$
F1Z2	$D\left(\dfrac{F1Z2\,\text{Force}}{2}\right),\ D\left(\dfrac{F1Z2\,\text{Time}}{2}\right)$
F1Z2-F1Z3	$D\left(\dfrac{F1Z2\,\text{Force}\cup F1Z3\,\text{Force}}{2}\right)-\left(\dfrac{F1\mathbf{Z}2\,\text{Force}}{2}\right)-\left(\dfrac{F1\mathbf{Z}3\,\text{Force}}{2}\right),$ $D\left(\dfrac{F1Z2\,\text{Time}\cup F1Z3\,\text{Time}}{2}\right)-\left(\dfrac{F1\mathbf{Z}2\text{Time}}{2}\right)-\left(\dfrac{F1Z3\,\text{Time}}{2}\right)$
F1Z2-F1Z4	$D\left(\dfrac{F1Z2\,\text{Force}\cup F1Z4\,\text{Force}}{2}\right)-\left(\dfrac{F1\mathbf{Z}2\,\text{Force}}{2}\right)-\left(\dfrac{F1\mathbf{Z}4\,\text{Force}}{2}\right),$ $D\left(\dfrac{F1Z2\,\text{Time}\cup F1Z4\,\text{Time}}{2}\right)-\left(\dfrac{F1\mathbf{Z}2\text{Time}}{2}\right)-\left(\dfrac{F1Z4\,\text{Time}}{2}\right)$
F1Z3	$D\left(\dfrac{F1Z3\,\text{Force}}{2}\right),\ D\left(\dfrac{F1Z3\,\text{Time}}{2}\right)$

(continued)

Table 4.1 (continued)

Spatial displacement features

Output signal(s)	Features
F1Z3-F1Z4	$D\left(\dfrac{F1\mathbf{Z}3\,\text{Force}\cup F1\mathbf{Z}4\,\text{Force}}{2}\right)-\left(\dfrac{F1\mathbf{Z}3\,\text{Force}}{2}\right)-\left(\dfrac{F1\mathbf{Z}4\,\text{Force}}{2}\right),$ $D\left(\dfrac{F1\mathbf{Z}3\,\text{Time}\cup F1\mathbf{Z}4\,\text{Time}}{2}\right)-\left(\dfrac{F1\mathbf{Z}3\,\text{Time}}{2}\right)-\left(\dfrac{F1\mathbf{Z}4\,\text{Time}}{2}\right)$
F1Z4	$D\left(\dfrac{F1\mathbf{Z}4\,\text{Force}}{2}\right),\ D\left(\dfrac{F1\mathbf{Z}4\,\text{Time}}{2}\right)$

Statistical features

Output signal	Measure	Features
F1X1	Force	Area, mean, standard deviation, norm
F1X2	Force	Area, mean, standard deviation, norm
F1Y1	Force	Area, mean, standard deviation, norm
F1Y2	Force	Area, mean, standard deviation, norm
F1Z1	Force	Area, mean, standard deviation, norm
F1Z2	Force	Area, mean, standard deviation, norm
F1Z3	Force	Area, mean, standard deviation, norm
F1Z4	Force	Area, mean, standard deviation, norm

The three sub-tables above demonstrate the *spatial* and *statistical* features examined in our research. Features measured in force refer to measurements in *Newton*, while time features refer to measurements in *seconds*. *Displacement* features were too numerous to display in this table and therefore were represented as sets using the binomial coefficient notation to express set membership. For example, set of distances between every two-feature combination from the four F1X1 Force point features would be represented as $D\left(\dfrac{F1X1\,\text{Force}}{2}\right)$

the GRF data series and iterate forward from that point until the values began to either decrease, when looking for a local maxima, or increase, when looking for a local minima. The point at which the values started to increase or decrease was returned as the local minima or maxima, respectively. This procedure proved problematic as incorrect extrema points were often returned due to noise in the data and footstep imperfections. To address this problem, the procedure was redesigned to accept two additional parameters: one to make the algorithm less sensitive to undesirable small peaks or troughs in the data, and another to "smooth" the data to more accurately determine the exact location of each extrema point.

The pseudo-code for the local maxima locator procedure is demonstrated in Fig. 4.2. The minima procedure is almost identical, but the inequality on line 5 is reversed. This procedure tracks the *maximum value* as the force values increase along the ridge, but when the values start to decrease rather than immediately returning the last maxima found, the procedure continues iterating until the *sensitivity threshold* is reached. If the sensitivity threshold is reached then the maximum

```
Input:      Initial GRF Index (X_init, Y_init),
            Integer threshold, Integer smoothing
            N = Sample Size
Output:     The Next Local Maxima (X_max, Y_max)

1   smooth(Y_point) ← Average(Y_point−smoothing/2 ··· Y_point + smoothing/2
2   Y_max ← Y_init
3   cnt ← 0
4   for Y_current = Y_smoothing ··· Y_N
5           if smooth(Y_current) > Y_max
6                   Y_max ← Y_current
7                   X_max ← X_current
8                   cnt ← 0
9           else
10                  cnt ← cnt + 1
11                  if cnt > threshold
12                          return (X_max, Y_max)
13                  end
14          end
15  end
16  return (X_max, Y_max)
```

Fig. 4.2 Local maxima finder pseudo-code. This figure demonstrates the pseudo-code for the algorithm we used for our local maxima locator

value recorded before the threshold counter began would be the largest value found so far and therefore would be returned as the local maxima. While the sensitivity threshold returned the greatest value as the local maxima, the value returned often turned out to be a single-record spike in force due to noise, rather than the actual visual top of the force ridge. To reduce the impact of noise, a previous gait-based study by Derawi et al. [7] imposed a weighted moving average on the data. For our research, a simple moving average was determined to be sufficient for noise reduction. When *data smoothing* is used, rather than looking for the single value maxima, the procedure looks for an averaged multi-value maxima and large spikes in the data are smoothed into a more level plane; this increases the likelihood that any maxima found by the procedure will indeed be the actual top of the ridge.

To locate the full set of point features on each output signal, the local minima and maxima locator procedures were *chained* together with the point located by the one locator forming the initialization index for the next in the chain. For example, in the Z-labeled output signals, the process would start at the first data record in the footstep then move forward through the records to locate the first local maximum. Next, the first local maximum would be assigned as the initialization index for the local minima locator, which, in turn, would locate the next local minimum. Finally, the *local minimum* returned by the local minima locator would be passed back into the local maxima locator as the initialization index, and the process would terminate with the locations of the first local maximum, the first local minimum, and the second local maximum all identified. When applied to the development dataset, this process was able to correctly locate the point features in all but one output signal in a single footstep sample, using *sensitivity threshold* parameter values in the range of 200–300 intervals for the Z/Y-labeled signals and 30 intervals for the X-labeled signal together with a *smoothing* parameter value of 5.

A visual inspection of the sample with the *missing point* features revealed that its F1Z2 output signal did not contain a maximum point where the expected first maximum would typically occur. Consequently, the minimum value point was also undefined. However, the graph, shown in Fig. 4.3, still demonstrated a well-defined *convex curvature* in the region where the first local maximum would be expected and *concave curvature* where the following local minimum would be expected. Thus, rather than leaving these features out, we decided to *estimate* the points of maximum convex and concave curvature, then use these points to approximate the expected positions of the local maxima and minima in the case that the initial locator process failed. Our estimation approach was based on a study by Castellanos et al. [3] that aimed to estimate the corner of the L-curve. In their work, they estimated that, given three points on a curve forming a triangle, the point of maximum curvature would be the position at which the middle of these three points

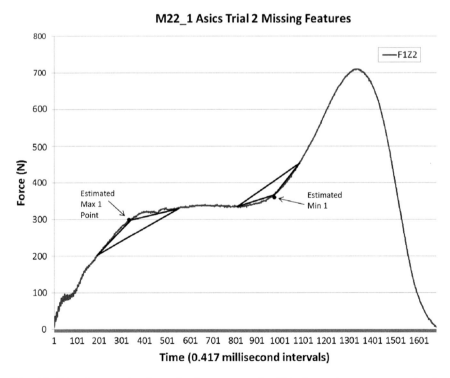

Fig. 4.3 Triangle approximation point locator example. This figure demonstrates a *triangle approximation* for finding missing point features. The points that maximized the area of the inscribed triangles were approximated as the expected position of missing local extrema features

had a *minimum angle value*. In our implementation we used a slightly different approach and searched for the three points, evenly spread across a 200 record-wide region of the graph, which *maximized the area* of the triangle formed by connecting each point. Again, the middle of these three points was taken as the approximate point of maximum curvature, as demonstrated in Fig. 4.3. The points that resulted from applying this technique to missing features-sample yielded a local maximum of (355, 307) and local minimum of (1012, 394). The estimated maximum point appeared relatively close to its average position of (368, 397) for the subject M22_1, while the estimated minimum point displayed significant error when compared to its average position of (736, 322). Because this missing feature showed up in only 1 of 300 data samples, the error was considered acceptable.

Having captured the full set of spatial point features for each sample, capturing remaining displacement and statistical features was relatively simple. The *displacements* were captured by finding the difference in time and force between the pairs formed by every combination of points, as listed in Table 4.1. The mean, Euclidean norm, and standard deviation were acquired by applying their respective statistical formulas to the set of all force recordings in the sample space, while the area under the curve was approximated by applying the *trapezoidal rule* to approximate the integral over the sample space (4.1).

$$\int_a^b f(x)dx \approx \frac{1}{2}\sum_{k=1}^{N}(x_{k+1}-x_k)(f(x_{k+1})+f(x_k)) \tag{4.1}$$

When the values for individual features in our 538 geometric feature space were compared, it became apparent that some features were much better discriminators than others. Furthermore, some features contained so much variability that using them for classification would likely lead to overfitting and decrease recognition performance. In [16], Rodríguez et al suggested an *optimization* technique to remove undesirable features from the geometric feature set. To accomplish this they used an exhaustive search process. The process started by identifying the stand-alone feature that produced the smallest EER during classification, then searched for the feature that, when combined with the initially discovered feature, produced the smallest feature pair EER. This process continued until the full feature set was sorted such that, when grouped with all the features ahead of it, each feature in the sorted set produced better performance than any of the features behind it. Using this process, Rodríguez's team found that, with only the 17 best of their initial 42 geometric features, they were able to reduce their EER by 22 %.

Building upon the work of [16], our research incorporated the *k-fold cross-validation* [19] into the original process. In our case, rather than sorting the features based on the best EER for a single training/testing subsets pair, the EER was calculated by taking the EER produced across all 10 possible training/testing cross-validation subsets in our development dataset. Our approach also differed in that we performed the optimization using a *weighted* KNN classifier (see Chap. 6), rather than the SVM classifier used in [16]. In this optimization, as well as all our other feature extraction optimizations and analysis, the K parameter was set to an arbitrary value of 5. The final optimization process functioned as follows:

(1) Let **O** be an empty set that will contain all geometric features ordered from most to least discriminative. Let **G** be the full geometric feature space.
(2) Take a feature from **G** that is not currently in **O** and add it to **O**.
(3) For each training/testing subset in the development dataset calculate the EER.
(4) Find the EER across every training/testing subset, and then remove the feature that was added in Step 2.
(5) Repeat Steps 2–4 for every feature not in **O**.
(6) Add the feature with the best averaged EER (Step 4) to **O** then repeat Step 2 until **O** contains every feature in **G**.
(7) Take the first N features that best reduce the EER and feature space dimensionality. These features will form the optimized geometric dataset.

The optimization process, when run on the original 538-feature geometric feature space using the development dataset, resulted in a significant improvement in recognition performance. The change in EER for the 110 initial optimization iterations is shown in Fig. 4.4. The diagram demonstrates a sharp drop in the EER up to the point when the optimal feature set contains 21 features, after which the EER flattens and even begins to increase as the optimal feature set grows in size. This behavior suggests that the classifier overfitting begins to hinder the recognition in

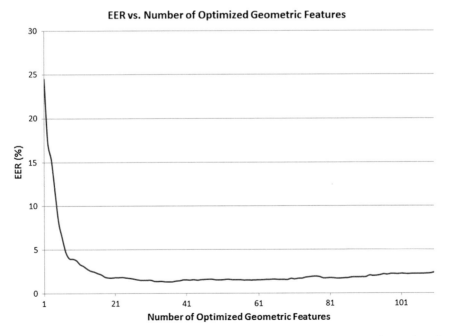

Fig. 4.4 EER versus number of optional geometric features. This graph compares the number of optimized features to the EER they produced. The count of features starts with the best performing single feature, then the best performing two features and continues with the features added resulting in smaller increases in performance as the size of the feature set grows

4.1 Geometric   63

large geometric feature spaces. Consequently, for the purpose of this book, the first 36 optimal features were chosen to form the *optimal geometric feature space*. This optimal feature space contained 6 *spatial* point features, 26 *displacement* features, and 4 *statistical* features, roughly corresponding to the frequency of each feature type in the original space. Furthermore, the measurement units for the optimal feature space consisted of about as many *time* features as there were *force* features. Compared with the original geometric feature space, the optimal feature space decreased the cross-validated development dataset EER from 8.48 % to 1.33333 %, equivalent to an 84 % increase in performance. Additionally, the 36-feature *optimized* geometric feature space represented a 93 % decrease in dimensionality over the *full* geometric feature space and a 99.7 % decrease over the roughly 12,800 record full footstep GRF data space (Fig. 4.5, Tables 4.2 and 4.3).

Table 4.2 Optimal geometric features

Optimal geometric features			
Feature	Unit	Feature	Unit
$D_{F1Z4MAX1\_F1X2MIN1}$	Time	$D_{F1Y1MAX1\_F1Y2MIN2}$	Time
F1X2 MIN1	Time	F1X2 MAX1	Force
$D_{F1Z2MAX2\_F1Y1MIN2}$	Time	F1Z3 MAX1	Force
$D_{F1Y1MIN2\_F1X1MAX1}$	Force	F1X1 MEAN	Force
F1Y2 NORM	Force	F1X1 MAX1	Time
$D_{F1X2MAX1\_F1X2MIN1}$	Force	$D_{F1Z3MAX1\_F1X2MAX1}$	Time
$D_{-F1Y2MIN1\_F1X2MAX1}$	Time	$D_{F1Y2MIN2\_F1X1MAX1}$	Force
$D_{F1Z3MAX2\_F1Y1MIN2}$	Time	$D_{F1Y1MIN1\_F1Y2MIN1}$	Force
$D_{F1Z1MAX1\_F1Z1MAX2}$	Force	$D_{F1Z3MIN1\_F1X2MIN1}$	Time
$D_{F1Z1MAX2\_F1Y1MIN2}$	Time	$D_{F1Z1MAX2\_F1Y2MIN2}$	Time
$D_{F1Z2MAX1\_F1Z2MAX2}$	Time	$D_{F1Z2MAX1\_F1Z4MAX2}$	Time
F1Y1 MIN1	Force	F1X1 AREA	Force
$D_{F1Z2MAX1\_F1Z2MAX2}$	Force		
F1Y1 MIN2	Force		
$D_{F1Y2MAX1\_F1X2MAX1}$	Force		
F1Y2 STDEV	Force		
$D_{F1Y1MIN2\_F1X2MAX1}$	Time		
$D_{F1Y1MIN2\_F1Y2MIN2}$	Force		
$D_{F1Y2MAX1\_F1X2MAX1}$	Time		
$D_{F1Z1MIN1\_F1X2MIN1}$	Time		
$D_{F1Y1MIN1\_F1X2MAX1}$	Time		
$D_{F1X1MAX1\_F1X2MAX1}$	Time		
$D_{F1Z1MIN1\_F1Z1MAX2}$	Force		
$D_{F1Z1MIN1\_F1X1MAX1}$	Time		

This table demonstrates the best 36 features in the feature set remaining after the geometric dataset was optimized. *Point features* are identified by the output signal name and extrema point type, while *displacement* features are identified with a 'D' and a subscript containing the two point features that formed the displacement. Each feature is categorized as either a measure of force (*Newton*) or time (*seconds*)

Table 4.3 Geometric feature extractor performance

Feature space comparison		
Feature space	Cross-validated EER (%)	Dimensions
Geometric	8.47777	538
Optimal geometric	1.33333	36

This table compares the performance of geometric feature spaces on the development dataset

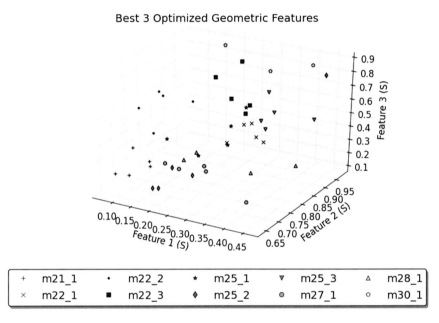

Fig. 4.5 Best three optimized geometric features. This diagram presents a visualization of the *three best* ranked optimized geometric features taken from footsteps belonging to *ten test subjects* and *projected* into a *3D frame*. *Five footsteps* per person are shown in this diagram and the footsteps belonging to each subject are distinguished by variations in the marker symbols used. For better visualization the range for each feature has been *standardized* as [0,1]

4.2 Holistic

Feature extraction techniques defined by heuristics, including the geometric technique described in the previous section, can be powerful data discriminators, but
rely on *expert knowledge* to identify the most valuable characteristics in a dataset.
When datasets are complex or have not been thoroughly studied, heuristic-based
techniques can suffer. The alternative to heuristic-based feature extraction
approaches involves using machine learning techniques to discover important data
characteristics. These techniques do the work of identifying the most statistically

significant characteristics in a dataset, according to some predefined learning model. For the purpose of this book, the non-heuristic techniques have been categorized into those that only perform feature extraction after a transformation of the data *domain* has occurred and those that only work with the data in its original domain. We refer to the latter category of feature extractor as *holistic* techniques following the usage of the term in [15, 16], and in our case the raw data being processed is represented in the *time domain*.

As noted in the previous chapter, several existing footstep GRF recognition studies [1, 11, 15, 16] have implemented holistic feature extraction solutions. In its simplest form, the holistic approach involves no feature extraction at all; instead entire data samples are passed *as-is* for classification and the determination of important characteristics is left to the classifier. Unfortunately, with each of our data samples consisting of approximately 12,800 records, passing the raw data to a classifier would be computationally undesirable and would likely lead to massive classifier overfitting and very poor recognition performance. To address this problem, we have based our holistic feature extraction on the dimensionality-reducing holistic technique proposed by Rodríguez et al. in [15]. In their work, they proposed a way to reduce the dimensionality of the original dataset using *principal component analysis* (PCA). Using PCA, Rodríguez's team discovered 80 features, from 4200 in their original dataset, which accounted for 96 % of the dataset's original information (*variance*). When running PCA on a dataset with multiple persons without a high level of variation between samples of a single individual, because variance represents variability in the dataset, we would expect to see the features with the greatest variance across the entire dataset also represent the features with the greatest disparity between individuals; the discovery of such distinguishing features would be important to achieve a strong recognition performance.

PCA involves transforming the original feature space, in our case, the approximately 12,800 records holistic sample space, into a new feature space where the new features, or *principal components* (PCs), are *uncorrelated* and ordered according to the amount of variance they represent [8]. The greatest challenge when implementing PCA comes from the need to generate a transformation that is able to represent the data in the PC feature space. Once this transformation has been generated, any given sample can be projected into the new PC feature space and dimensionality and can be reduced by extracting only the small group of PCs that account for the majority of the original feature space's variance. A five-step process to generate the needed PCA transform is described by Lindsey Smith [17]. The process begins by calculating the *mean* for each feature in the training subset, and then proceeds to subtract the calculated means from the respective features in each training sample; the end result is a dataset whose *mean is zero*. Next, the *covariance matrix* for the zero-mean dataset is calculated according to (4.3), with A representing the $n \times n$ covariance matrix and Dim_x representing the xth dimension of the sample space. In this calculation, the matrix entries of A are formed by taking the covariance calculation (4.2) between each dimension for all the N samples across the two dimension vectors parameters.

$$cov(X, Y) = \sum_{i=1}^{N} \frac{(x_i - \bar{x})(y_i - \bar{y})}{N} \tag{4.2}$$

$$A^{n \times n} = (c_{i,j}, c_{i,j} = cov(\text{Dim}_i, \text{Dim}_j)) \tag{4.3}$$

Once we have the covariance matrix, we need to calculate its *eigenvectors* and *eigenvalues*; these can be found by first solving for the eigenvalues (4.5) then substituting each eigenvalue back into (4.4) to solve for the eigenvectors. In calculating the eigenvalues, we are looking for the $n \times 1$ eigenvalue vector λ that, when multiplied against the $n \times n$ identity matrix I and subtracted from our $n \times n$ covariance matrix A, produces a determinant value of 0 when eigenvectors are present. With this understanding we can generate a *characteristic polynomial equation* to solve for the eigenvalues and substitute the result back into (4.4) to solve for the eigenvectors. To find the eigenvectors, we must find all values of x that, when multiplied with the covariance matrix A, result in some scalar multiple λ of x. Having already found the eigenvalues λ, the vector solutions x form the eigenvectors and, in the $n \times n$ covariance matrix eigenvectors, there will be n of them [17].

$$Ax = \lambda x \tag{4.4}$$

$$\det(A - \lambda I) = 0 \tag{4.5}$$

When all eigenvalues and eigenvectors have been calculated, we will find that the *magnitude* of the eigenvalues corresponds to their respective degrees of variance in the original dataset; with this knowledge, we can extract a small set of eigenvectors that accounts for the majority of the variance in the original dataset to form our PCs. Finally, when we want to project the data samples from our original feature space into the new smaller PC feature space, we can do so using this small set of eigenvectors and applying (4.6). In this equation, X is a $1 \times n$ input vector to be projected into the PC-reduced feature space of size m and W is the $n \times m$ PC-reduced eigenvector matrix, which when *transposed* contains the eigenvectors in its rows. Under this equation, each feature in X is assumed to have had its respective *training subset-calculated* mean value subtracted from it, so when X^T and W^T are multiplied the result is Y, an $m \times 1$ *projection* of X into the *reduced feature space*.

$$Y = W^T \times X^T \tag{4.6}$$

The implementation of the PCA feature extraction technique in our biometric system was accomplished through integration with the *Accord.NET* PCA library developed by César Souza [6]. This library featured several improvements over the process described in the previous paragraphs. For instance, rather than calculating the eigenvalues and eigenvectors directly, a task that can be computationally intensive, Souza acquired both sets using the computationally efficient and more flexible *singular value decomposition* (SVD) algorithm [9]. Souza's library also gave the option of using a *correlation matrix* instead of the covariance matrix when generating the PCs; an important option because the use of the correlation matrix can generate a better performance when features have broad differences in their variances.

Before we could run PCA on our development dataset, we first needed to *standardize* the size of each data sample. Variations in stepping speed meant sample length and therefore the number of features per sample did not match up across the dataset. This was problematic for PCA, which expects a standard number of features to be present to both generate and carry out the feature space transform. To address this problem, we initially followed the approach taken in [16], where an arbitrary number of records, large enough to represent full footsteps, were captured for each output signal. For our dataset, we determined that 2000 records per output signal, for a total of 16,000 records per sample, were sufficient to obtain all information of value in any given footstep. To ensure that each sample contained the same number of records, extra *zero-valued* records were appended to signals with less than 2000 records. We refer to this standardization approach as the *point-based* holistic approach.

Having established the process to perform PCA, when we wanted to perform classification using the dimensionality-reduced holistic feature extraction technique we did so through the following steps:

(1)   Perform PCA on the training data subset (i.e., five samples per person for all enrolled persons), and pick the best PCs to form the dimensionality-reducing feature space transform.
(2)   Project the training data subset into the new PC feature space.
(3)   Train the chosen classifier using the transformed training data samples.
(4)   Project a data sample from the testing subset into the new PC feature space.
(5)   Perform classification on the transformed testing data sample using the classifier from Step 3.

The results from running the point-based holistic feature extraction technique on our KNN classifier are shown in Fig. 4.6. To better assess the accuracy of the point-based holistic approach, *cross-validation* was performed with a *unique* PCA transform generated for every training/testing subset; the results in Fig. 4.6 reflect the average performance achieved across the 10 iterations of cross-validation needed to cover our entire development dataset. The best point-based holistic performance was found to occur when using the covariance matrix during PCA, and the optimal point-based holistic feature set contained the first 15 PCs accounting for approximately 98.7 % of the dataset variance with an EER of 3.42 %.

In addition to the point-based holistic approach, during our research, we developed a new approach for standardizing the size of the footstep GRF sample space passed in for PCA. This new approach, which we refer to as the *area-based* holistic approach, involved dividing each sample into a standard number of

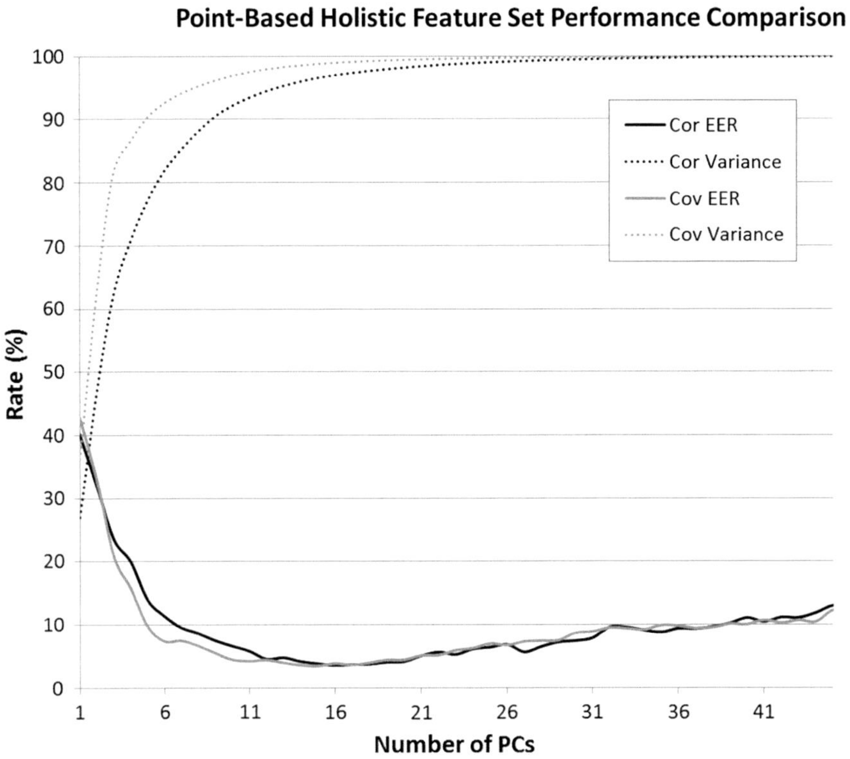

Point-Based Holistic Feature Set Performance Comparison

Fig. 4.6 Point-based holistic feature space performance comparison. This figure demonstrates the performance of our *point-based* holistic approach using both *covariance* (cov) and *correlation* (cor) matrices during PCA. The PC set size is shown as a function of the EER it produces and the approximate dataset variance it represents

temporally equal *proportional regions* and forming a new dataset with the set of regional areas. The graph in Fig. 4.7 shows how this would work on a single output signal with the dataset divided into 8 regional areas; in practice, we would want far more regions for better resolution when performing PCA.

To begin the process of generating our new area-based sample space, we defined our original data sample (**D**), of dimensionality (*size*) N, as a series of records with the value t_i representing the time the sample interval i occurred at, and f_i the force value at t_i, as demonstrated in (4.7).

$$D = ((t_0, f_0), (t_1, f_1), \ldots (t_{N-1}, f_{N-1}))$$ (4.7)

$$S(x) = \frac{N}{M} \times x$$ (4.8)

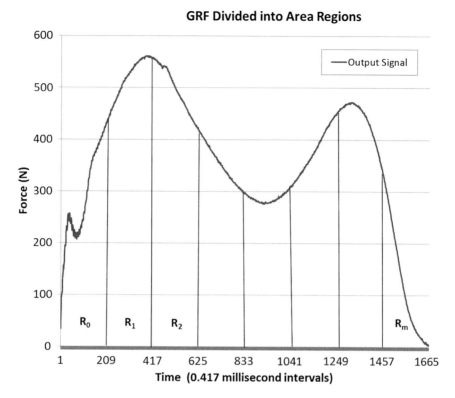

Fig. 4.7 Footstep GRF divided into area regions. This figure demonstrates an example of the GRF data from one of our output signals divided into eight regions. Using our *area-based* approach, the areas for each region would be calculated and stored as a new sample space

To support our calculations, we defined the function (4.8) to mark the start of an area region x with respect to our original space; with N being the dimensionality of the original sample space (D) and M being the dimensionality of the new area-based sample space.

Using the definitions in (4.7) and (4.8), the area, R_i, for a given region i in the new area-based space can then be calculated using the sum of a set of *trapezoidal approximations* as shown in (4.9). Because the area regions may start and/or end in the space between the intervals of the original sample, the equation was divided into *three parts*: the first part finds the approximate area for any partial region preceding the first original dataset record (t, f) in R_i, the second part calculates the approximate area for the set of original dataset records falling within R_i, and final part approximates the area for any partial region following the last original dataset record in R_i.

$$R_i = \frac{1}{2}\left(\lceil S(i)\rceil - S(i)\right)\left(t_{\lceil S(i)\rceil} - t_{\lceil S(i)\rceil - 1}\right)\left(f_{\lceil S(i)\rceil} + f_{\lceil S(i)\rceil - 1}\right)$$

$$+ \frac{1}{2}\sum_{j=\lceil S(i)\rceil}^{\lfloor S(i+1)\rfloor - 1}\left(t_{j+1} - t_j\right)\left(f_{j+1} + f_j\right)$$

$$+ \frac{1}{2}\left(\lfloor S(i+1)\rfloor - S(i+1)\right)\left(t_{\lfloor S(i+1)\rfloor + 1} - t_{\lfloor S(i+1)\rfloor}\right)\left(f_{\lfloor S(i+1)\rfloor + 1} + f_{\lfloor S(i+1)\rfloor}\right)$$

$$(4.9)$$

Finally, to get the complete area-based sample space, we simply calculate R_i for $i = 1:M$ as shown in (4.10). In performing this calculation across a full dataset, it is important to note that for us to achieve a *standard sample size* for all available samples, the number of *area regions* (M dimensions) must be smaller than the number of *points* in the smallest expected original sample (N dimensions); in our dataset, no samples were less than 1400 records in length, giving us a well-defined boundary from which to work.

$$R = (R_0, R_1, \ldots R_{M-1}), \quad M < N \qquad (4.10)$$

As noted in the previous paragraph, to use the area-based holistic approach, we needed to decide on the number of area regions into which the original sample space should be divided. After testing the performance over 50-region intervals, from between 1000 regions per output signal and 50 regions per output signal, we found that the number of regions had relatively little impact on the performance and settled on *500 regions* per signal for a new total sample space of 4000 area features. The results from running the area-based holistic feature extraction technique on our KNN classifier are shown in Fig. 4.8. As was the case when assessing the point-based feature extraction technique, the results shown here were also obtained via cross-validation. In contrast to the results obtained from the point-based holistic technique, when the *area-based* holistic technique was used there was a significant difference in performance between the covariance matrix and the correlation matrix-based approaches, with the covariance-trained system achieving an optimal EER of 3.0 % and the correlation achieving an optimal EER of 2.55 % (Fig. 4.9, Table 4.4).

Comparing the point-based holistic approach to the area-based holistic approach, we found that both approaches resulted in a dimensionality reduction of about 99.8 %. We also found that the area-based approach achieved better recognition performance; however, neither approach performed as well as the optimized geometric approach in the previous section. This may be a consequence of the variations in stepping speed between samples. Differences in the stepping speed meant the sample space features passed for analysis via PCA were not being assessed on a single *standard scale*, but rather on *multiple scales* depending on the step duration. In [15, 16], the point-based holistic feature extraction approach achieved a better performance than the optimized geometric approach. Part of this may be due to the

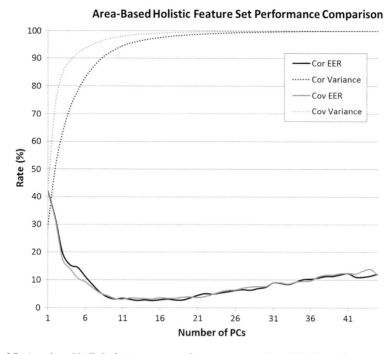

Fig. 4.8 Area-based holistic feature space performance comparison. This figure demonstrates the performance of the *area-based* holistic approach using both *covariance* (cov) and *correlation* (cor) matrices during PCA

fact that these studies used a much larger training dataset than ours to generate their PC features. But these studies also used a data *normalization* technique to ensure that the different samples were compared on the same scale. To address that issue, we will later explore various normalization techniques in Chap. 5.

Table 4.4 Holistic feature extractor performance

Feature space comparison		
Feature space	Cross-validated EER (%)	Dimensions
Point-based holistic (covariance)	3.42222	15
Point-based holistic (correlation)	3.54444	16
Area-based holistic (covariance)	3	11
Area-based holistic (correlation)	2.55555	15

This table compares the performance of holistic feature spaces on the development dataset

Best 3 Holistic Features

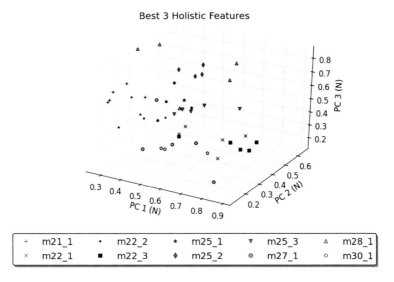

| + | m21_1 | • | m22_2 | ★ | m25_1 | ▼ | m25_3 | △ | m28_1 |
| × | m22_1 | ■ | m22_3 | ◆ | m25_2 | ⊙ | m27_1 | ○ | m30_1 |

Fig. 4.9 Best three holistic features. This diagram presents a visualization of the 3 *best area-based* holistic PC features taken from footsteps belonging to *ten test subjects* and *projected* into a *3D frame*. *Five footsteps* per person are shown in this diagram and the footsteps belonging to each subject are distinguished by variations in the marker symbols used. For better visualization the range for each feature has been *standardized* as [0,1]

4.3 Spectral

Important features are not always apparent in the time domain and occasionally may become more apparent when a data sample is analyzed in the *frequency domain*. Two previous footstep GRF recognition studies [4, 18] have suggested techniques for extracting features from the frequency domain; we refer to these features as *spectral features*. In [18], Suutala and Röning proposed that features be extracted from the *magnitude* of the GRF and GRF *derivative frequency spectra*; while in [4], Cattin proposed that features be extracted from another representation of the magnitude spectra called the *power spectral density* (PSD), with only the derivative of the vertical GRF component examined. Both studies used PCA-based dimensionality reduction techniques to assist in the discovery of a smaller optimized spectral feature space. In our research, we decided to test both the magnitude spectra and PSD approaches, while also incorporating our PCA technique from the previous section.

To apply the two spectral approaches and accurately assess their effectiveness, we extended our holistic approach to include two more steps. Prior to performing PCA, we have added a *filter* which converts the dataset to the frequency domain then returned either its magnitude spectra or PSD. Additionally, to stay consistent with the work presented in the previous GRF spectral feature studies, we have included an optional filter to obtain the GRF derivative before the transformation to

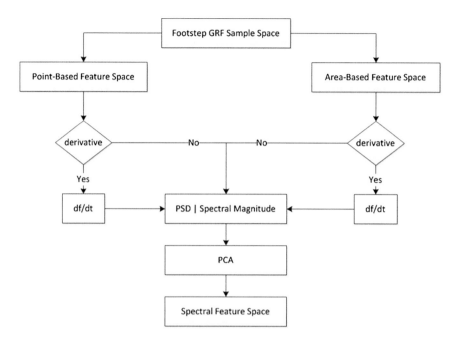

Fig. 4.10 Spectral feature space generation process. This figure represents the process used to generate a spectral feature space. Taking the derivative of the GRF before processing is optional, but the method chosen for training PCA must also be used for testing

the frequency domain. Additionally, as was the case for the holistic feature extraction, the process begins by generating either a standardized *point-based* sample space or one based on regional *areas* of the GRF. The resulting spectral features extraction process is shown in Fig. 4.10, while an example comparing the GRF to the GRF derivative is shown in Fig. 4.11.

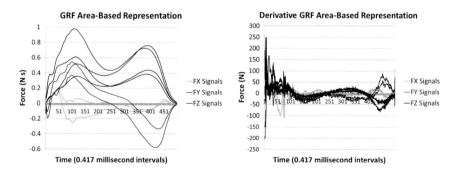

Fig. 4.11 Footstep GRF versus derivative. The graph on the *left* of this figure represents the regional *area-based* features in *Newton-seconds* at the time interval they were measured, while the graph on the *right* represents its *derivative*

Transforming the finite data series to the frequency domain is accomplished using the *discrete Fourier transform* (DFT) (4.11), with c_j in the equation containing the GRF or derivative GRF record at the jth interval (dimension), k being the index of the frequency spectral lines, and N representing the sample size (dimensionality). When this transform is performed, it returns a series of *complex numbers* (C_k) that can be processed to determine the spectral *phase* and *magnitude*.

$$C_k = \sum_{j=0}^{N-1} c_j e^{\frac{2\pi ijk}{N}} \quad k = 0,\ldots,N-1 \tag{4.11}$$

We find the spectral magnitude of this series by calculating the *square root of the sum of the squares* for each of its real and imaginary parts (4.12). In this equation, we demonstrate the calculation of the frequency spectrum magnitude at index k, with *Re* representing its C_k's real term and *Im* its complex term.

$$|C_k| = \sqrt{Re(C_k)^2 + Im(C_k)^2} \tag{4.12}$$

Calculating the PSD is a little more challenging and requires finding an estimation called a *periodogram*. For the purpose of this book, we have implemented our periodogram and DFT using the code presented by Press et al. in Numerical Recipes [14]. In their application, they used a *fast Fourier transform* (FFT) to optimize the derivation of the frequency domain and included several optional non-rectangular window functions to counter *spectral leakage* during the calculation of the periodogram. Spectral leakage becomes a problem when a signal does not end at its periodic interval, which results in the unwanted "leakage" of any incomplete periodic cycles at the signal's boundary into nearby frequency bins. In our sample space, leakage was not found to be a major issue, so in our implementation, we opted for the *rectangular-windowed* calculation of the periodogram. This periodogram estimation of the power spectrum at $N/2 + 1$, as defined by [14], is demonstrated in (4.13). In this equation, f_k is defined only for *zero* and *positive frequencies* and f_c is the *Nyquist critical frequency*, as defined in (4.14) for the interval Δ.

$$P(0) = P(f_c) = \frac{1}{N^2}|C_0|^2$$

$$P(f_k) = \frac{1}{N^2}\left[|C_k|^2 + |C_{N-k}|^2\right], \quad k = 1, 2, \ldots \left(\frac{N}{2} - 1\right)$$

$$P(f_c) = P\left(f_{\frac{N}{2}}\right) = \frac{1}{N^2}\left|C_{\frac{N}{2}}\right|^2 \tag{4.13}$$

$$f_k \equiv \frac{k}{N\Delta} = 2f_c\frac{k}{N} \quad k = 0, 1, \ldots \frac{N}{2}$$

$$f_c \equiv \frac{1}{2\Delta} \tag{4.14}$$

The spectral magnitude and PSD that resulted from transforming an area-based sample footstep GRF and its derivative are shown in Figs. 4.12 and 4.13, respectively. As demonstrated in these figures, our spectral sample space is derived by performing the frequency domain transformation *separately* on each output signal rather than on any combination of signals; however, when PCA is performed, all eight extracted frequency domains are *concatenated* into a single space for analysis. The aforementioned frequency domain representations showed little in terms of visual characteristics since they tended to be heavily dominated by the strength of a small number of lower frequency bins. Yet, the performance of these spectral feature sets, shown in Figs. 4.14 and 4.15, turned out to be similar to the performance achieved using the holistic extraction technique.

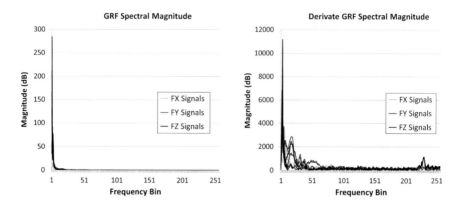

Fig. 4.12 Footstep GRF spectral magnitude. This figure provides a comparison of the *spectral magnitude* between the *area-based* GRF and *area-based derivative* GRF using the sample in Fig. 4.11

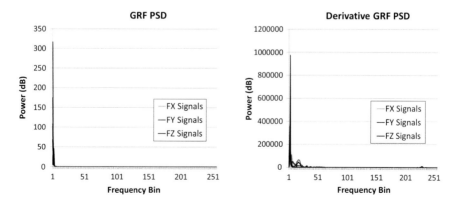

Fig. 4.13 Footstep GRF power spectral density. This figure provides a comparison of the PSD between the *area-based* GRF and *area-based derivative* GRF using the sample in Fig. 4.11

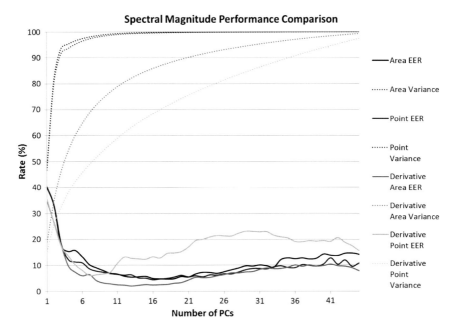

Fig. 4.14 Spectral magnitude performance comparison. This figure compares the performance of our four spectral magnitude feature spaces

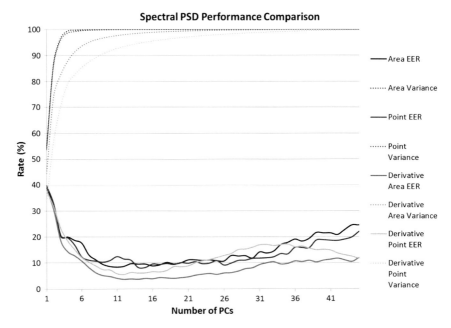

Fig. 4.15 Spectral PSD performance comparison. This figure compares the performance of our four spectral PSD feature spaces

The combination of the area/point-based approaches and derivative/non-derivative representations gave us four different spectral feature spaces to analyze. The results from performing a classification on our development dataset with these four spectral feature spaces in terms of both spectral magnitude and PSD are shown in Figs. 4.14 and 4.15, respectively. All demonstrated results were achieved using the *covariance* PCA configuration, which performed better than the correlation configuration. Once again, we used our KNN classifier to acquire the EER, with an area-based space composed of 500 area regions per output signal and point-based space composed of 2000 points per output signal. In both the spectral magnitude and PSD feature spaces, the best performance came from the *area-based GRF derivative* features. Furthermore, while each extractor reduced the feature space dimensionality by 99 %, the spectral magnitude feature spaces clearly performed better than the PSD feature spaces, with an optimal spectral magnitude EER of 2.02 % versus an optimal PSD EER of 3.68 % (Fig. 4.16, Table 4.5). However, it should be noted that while our analysis found that features extracted from the spectral magnitude sample space performed better than those extracted from the PSD feature space for a single footstep, a direct comparison could not be made with the PSD method used in [4] since the work in that paper performed recognition using a *multi-footstep* sample space and a *generalized* variation of PCA.

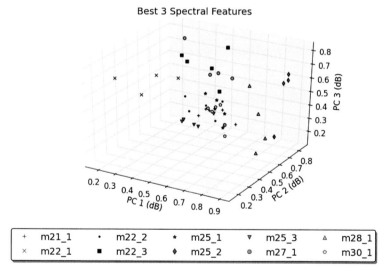

Best 3 Spectral Features

| + | m21_1 | · | m22_2 | * | m25_1 | ▼ | m25_3 | ▲ | m28_1 |
| × | m22_1 | ■ | m22_3 | ♦ | m25_2 | ⊚ | m27_1 | ○ | m30_1 |

Fig. 4.16 Best three spectral features. This diagram presents a visualization of the *spectral magnitude* PC features, obtained using the *area-based derivative* sample space taken from footsteps belonging to *ten test subjects* and projected into a *3D frame*. *Five footsteps* per person are shown in this diagram and the footsteps belonging to each subject are distinguished by variations in the marker symbols used. For better visualization the range for each feature has been *standardized* as [0, 1]

Table 4.5 Spectral feature extractor performance

Feature space comparison		
Feature space	Cross-validated EER (%)	Dimensions
Point-based spectral magnitude	4.37777	16
Point-based derivative spectral magnitude	6.1	7
Area-based spectral magnitude	4.67777	18
Area-based derivative spectral magnitude	2.02222	13
Point-based spectral PSD	9.4	19
Point-based derivative spectral PSD	5.55555	12
Area-based spectral PSD	8.38888	11
Area-based derivative spectral PSD	3.68888	12

This table compares the performance of spectral feature spaces on the development dataset

4.4 Wavelet Packet

In the previous section, we explored the idea that important GRF characteristics may be found in the frequency domain. The analysis of the frequency given a time domain-based dataset has traditionally been done via the application of the *Fourier transform* (the approach used in our spectral feature extractors). However, when the Fourier transform is applied, significant information regarding the *location* of particular frequencies will be lost [2]. In [12], Moustakidis et al. proposed an alternative form of GRF frequency analysis based on the *wavelet packet* (WP) transform. While the domain obtained by the Fourier transform is characterized by basis functions consisting of *sine* and *cosine* functions, the domain obtained by the WP transform is characterized by basis functions that are *localized* over a finite space and called *wavelets*. This space-localization property of the WP transform makes it possible to effectively analyze the frequencies that occur over a particular period of time; consequently, the domain resulting from the WP transform is often referred to as the *time–frequency domain*. In the research presented in this book, we refer to the features extracted in the WP time–frequency domain as *wavelet features*, and we have based our analysis of these features on the work done in [10, 12].

The feature extraction technique described in [12] was based on a proposal by Li et al. [10] to improve the classification of biomedical signals. In their proposal, they outlined a *two stage* process for extracting the features; the first stage involved performing a *WP transform*, while the second stage used *fuzzy sets* to identify the most discriminant features in the new WP space. The application of the fuzzy set-based feature identification technique was significant. Unlike the PCA-based approaches, which identified important characteristics with no prior knowledge of the subjects used in training, the fuzzy set technique was made fully aware of the subjects attached to each training sample and used this knowledge to construct a ranking of discriminative features. In machine learning, these class-aware algorithms are known as *supervised* learning models, and, by including this approach in our research, we were able to compare and contrast its performance with that of the *unsupervised* PCA approach; albeit, the comparison is done across differing domains.

As mentioned in the previous paragraph, the fuzzy WP feature extraction technique is divided into two stages. The first stage involves performing *wavelet packet decomposition* (WPD) on the sample space for each of our training samples. During WPD, each studied sample is passed through a *filter bank* defined by a chosen *wavelet function*; this filter bank consists of a *high* and *low pass filter*, and divides the sample space along the *center* of its frequency spectrum producing one subspace representing the upper half of the original sample space's frequency spectrum and the other representing its lower half. Next, the resulting *frequency subspaces* will each be passed back through the wavelet filter bank producing four subspaces, and this process of further dividing the frequency subspaces will continue until a specified level of decomposition is completed or the Nyquist limit for the sample space is reached. The end result of the WPD can be represented as a tree with exactly two nodes per branch.

In the second stage of the fuzzy WP feature extraction technique, we search the WPD tree for its most discriminative characteristics. To do this, we first want to find the most discriminative set of WPD *nodes* covering the entire sample space frequency spectrum such that there is no overlap between the spectra of individual nodes. This is done to ensure that no redundant information is used in analysis and this representation is called the *optimal WPD*. To find the optimal WPD, we need to rank each WPD node for its discriminative ability. In [10], node ranking was accomplished using a function based on the *fuzzy c-means clustering approach*. For each WPD node, this function examines all training samples and determines a degree to which the WPD coefficients in the given node correspond to their mean value for the subject they represent; these values are summed up for all coefficients in the node and the higher the resulting total, the better the discriminative ranking assigned to the given node.

In Eq. (4.15) the derivation of the *fuzzy membership criterion* ($F(X)$) used to rank the nodes in the WPD tree, is demonstrated, with X representing the feature space for a single node, c representing the number of classes (subjects), and A_i the set of training data samples belonging to class i. In this equation, numerical scores are generated by u_{ik}, the *fuzzy c-means objective function* (4.16), which determines the degree to which the wavelet coefficient vector for the training sample k (corresponding to x_k') in node X belongs to the class i. In this function, the 'prime' symbol represents the vectors reduced using an *exclusion criterion* (4.17) (which will be discussed in the next paragraph) and σ represents the normalization of coefficients by their standard deviation; also, v_i' represents the vector containing the wavelet coefficient means of class i, and b the *fuzzifier* that modifies the *shape* of the membership function. For the purpose of our research we set $b = 2$, and applied the two *boundary conditions* identified in [10]: if $x_k' = v_i'$, then $u_{ik} = 1$, and, if $x_k' = v_j'$, $i \neq j$, then $u_{ik} = 0$.

$$F(X) = \sum_{i=1}^{c} \sum_{k \in A_i} u_{ik} \tag{4.15}$$

$$u_{ik} = \left[\sum_{j=1}^{c} \left(\frac{\left\| x_k' - v_i' \right\|_{\sigma'}^2}{\left\| x_k' - v_j' \right\|_{\sigma'}^2} \right)^{\frac{1}{b-1}} \right]^{-1} \tag{4.16}$$

An unfortunate consequence of the unrestrained optimal WPD process is that it can suffer when the dataset contains coefficients that produce poor degrees of membership; in this case, a single poor coefficient or small group of *poor coefficients* may contribute to a significant reduction in node ranking for nodes that contain strong discriminators and otherwise would rank strongly. To counter this condition, Li's team proposed the application of *exclusion criterion* (4.17) to remove samples with poor discriminative ability prior to calculating the optimal WPD. Additionally, to prevent any single coefficient from having an undesirably large or small impact on the node rankings Li's team normalized each coefficient by its standard deviation. In Eq. (4.17), for any given feature j, the exclusion criterion $(D(j))$ is calculated by taking the maximum distance in *mean* values (v_{ij}) for the given feature and class $i = 1...M$, dividing it by twice the standard deviation of the given feature, then comparing it against the *retention* threshold r. If $D(j) < r$, then the feature j is excluded from the feature space.

$$D(j) = \frac{\max\left\{ v_{ij} \big|_{i=1}^{M} \right\} - \min\left\{ v_{ij} \big|_{i=1}^{M} \right\}}{2\sigma_j} < r \tag{4.17}$$

For our research, we have used the optimal WPD technique described above, but also included the additional step of transforming the original sample space into either a dimensionally standardized *point-based* or *area-based* sample space (see Sect. 4.2) prior to performing the WPD. We have also differentiated our GRF wavelet feature research from the work done in [12] by performing the WPD with two previously untested wavelet functions, Legendre 04 (lege04) and Legendre 06 (lege06); the performance of these two wavelet functions was compared with the two best wavelet functions in [12], the Coiflet 06 (coif06) and Daubechies 04 (daub04). Fig 4.17 illustrates an optimal WPD for the F1Z4 signal, calculated using the area-based dimension standardization approach with a *coif06* wavelet function.

Finding the optimal WPD gives us the wavelet feature space, but the application of dimensionality reduction to this space requires an additional step. In [10, 12], the dimensionality reduction was accomplished by calculating the *fuzzy membership* for each individual wavelet feature, *sorting* the feature indices by the membership values returned, then forming the dimensionally reduced feature set as the first N indices, where N is less than the total number of dimensions. The *fuzzy membership feature rank scoring function* $(F(j))$ used to accomplish this for any given feature j is demonstrated in (4.18). In this equation, i is taken to represent the class,

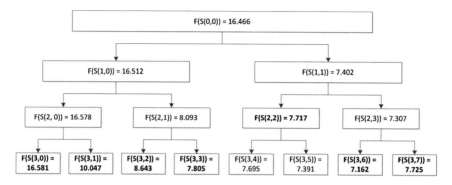

Fig. 4.17 Optimal wavelet packet decomposition. The optimal WPD tree for the F1Z4 output signal. Subspaces are represented as *S(j, k)*, where *j* represents the decomposition level and *k* spectral position of the node. The *score ranking* for each node is shown and the *optimal decomposition* is *highlighted in bold*

c the total number of classes, A_i the training samples belonging to class *i*, x_{ki} the feature *j* in sample *k*, v_{ij} the mean for feature *j* in class *i*, and u_{ij} the membership score for feature *j* in class *i*.

$$u_{ij}(x_{kj}) = \left[\sum_{m=1}^{c} \frac{(x_{kj} - v_{ij})^2}{(x_{kj} - v_{mj})^2} \right]^{-1}$$

$$F(j) = \sum_{i=1}^{c} \sum_{k \in A_i} u_{ij}(x_{kj})$$

(4.18)

The described membership function reflects on the *discriminatory power* for each feature and therefore the features at the start of the sorted list should be the best suited for classification. To find the best set of features for classification, the choice of *N* can be optimized in the same way that the number of PCs for holistic features was optimized, by plotting the size of the feature set against the EER it produces.

When applying the described WP feature extraction technique to the GRF data, Moustakidis's team calculated the optimal WPD for each of the GRF components separately, then performed the feature ranking on the space derived from combining every optimal WPD. We used a similar approach in our wavelet feature extractor, but rather than calculating the optimal WPD separately for each GRF component, we calculated it separately for each of our eight GRF output signals. Using this approach, the incorporation of the WP feature extraction technique into the person classification was achieved through the following steps:

(1) Standardize the dimensionality of each subset using the area-based or point-based approaches in Sect. 4.2. Then compute the WPD for each output signal in the training data subset.

(2) Use the WPDs calculated in Step 1, together with (4.15), (4.16) and (4.17), to
     calculate the optimal WPD; then transform the output signals of each training
     data sample into their respective optimal WPD space.
(3) Combine the optimal WPDs for each output signal to form the full wavelet
     feature space and then use (4.18) to apply a fuzzy membership rank to each
     feature.
(4) Sort the features by rank and take the first N best features to form the new
     dimensionally reduced wavelet feature space.
(5) Transform each training data sample into the reduced wavelet feature space.
(6) Train the classifier using the transformed training samples.
(7) Transform a data sample from the testing subset into the reduced wavelet
     feature space.
(8) Perform classification on the transformed testing data sample using the clas-
     sifier from Step 6.

   To implement the WP feature extractor, we first converted the Christian
Scheiblich's *JWave* [5] into C# and used it to perform the WPD, then integrated it
with our own fuzzy membership-based C# solution for finding the optimal WPD
and wavelet features. To find the reduction in dimensionality that best optimizes the
WP feature extractor's performance, we plotted the 100 best features for various
extractor configurations against the EER produced as each feature was successively

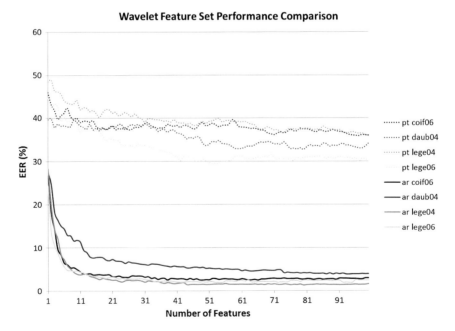

Fig. 4.18 Wavelet feature set performance comparison. This figure demonstrates a comparison of
the performance between different wavelet feature extractor configurations. In this diagram 'pt'
represents *point-based* approaches while 'ar' represents *area-based* approaches

included in the feature space; optimal feature sets larger than 100 dimensions were not considered competitive with the alternative feature extraction techniques and thus ignored. Once again, the classification results were measured using our KNN classifier and cross-validation was performed to improve the accuracy. To remove the *bias*, in each cross-validation iteration a new WP feature extractor was generated. Our tested configurations included *four wavelet functions* (coif06, daub04, lege04, and lege06), the wavelet decomposition *depth* ($L = 4$) and *retention threshold* ($r = 0.3$) used in [12], as well as both our *point-* and *area-based* dimension standardization approaches. The point-based standardization (2048 dimensions) and area-based standardization (512 dimensions) were set as powers of 2 to facilitate any level of decomposition. The results from running these wavelet packet extractor configurations are shown in Fig. 4.18.

Analyzing the results in Fig. 4.18, it is apparent that the WP feature extractor performed much worse when run on the *point-based* sample spaces. This may be a consequence of the different used sample space sizes; while the point-based approach produced a sample space of 16,384 dimensions for reduction, the area-based approach consisted of only 4096 dimensions. Given such a large number of point-based dimensions, the extractor may have assigned a *high ranking* to a

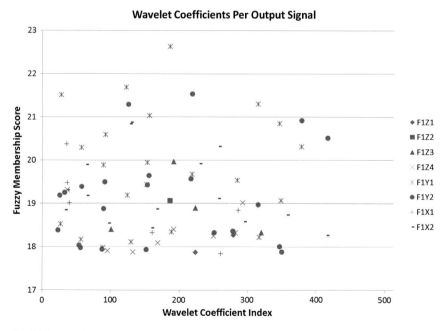

Fig. 4.19 Wavelet coefficients per output signal. This figure demonstrates the 80 best features derived using the *area-based* dimensionality *standardization* approach with a WP feature extractor based on the *lege04* wavelet function. The *fuzzy membership score* in the y-axis refers to the feature ranking scores returned by the equation (4.18), while the *wavelet coefficient index* refers to the index of the feature in the optimal WPD feature space

number of features that performed well individually, but shared redundant information with their high ranking peers and contributed to no increase in performance when *grouped* into a feature set. Another possibility may be that the WP feature extraction technique is not well suited for identifying the GRF features when substantial differences in the feature space *alignment* are present, as was the case for the point-based approaches. In [12], samples were discretized into 700 dimensions per GRF component, producing a total of 2100 dimensions and a proportional sample space similar to the one produced by our area-based approach. Classification performance in [12] was relatively close to the performance achieved using our area-based WP feature extractor, yet, unlike the results of [12], our best performance came when WPD was run using the *Legendre 04* wavelet function.

Using the area-based WP feature extractor with the Legendre 04 wavelet function, we achieved a best EER of 1.28 % for a feature set of 80 dimensions (a 99.3 % decrease in dimensionality (Fig. 4.20, Table 4.6)). One interesting by-product of the WP feature extractor was the grouping of features according to their corresponding *output signals*. This is demonstrated in Fig. 4.19, where the 80 best features are labeled with their output signal and measured by their fuzzy membership ranking and position in the wavelet feature space. We discovered that, in the first cross-validation of our development dataset, 45 of our best wavelet features were derived from the *anterior–posterior* GRF component, 14 were

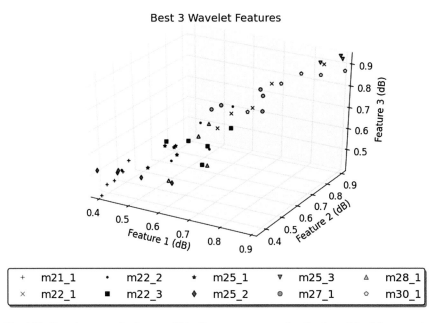

Fig. 4.20 Best three wavelet features. This diagram presents a visualization of the *3 best wavelet features*, obtained using the area-based sample space taken from footsteps belonging to *ten test subjects* and *projected* into a *3D frame. Five footsteps* per person are shown in this diagram and the footsteps belonging to each subject are distinguished by variations in the marker symbols used. For better visualization the range for each feature has been *standardized* as [0, 1]

Table 4.6 Wavelet feature extractor performance

Feature space comparison		
Feature space	Cross-validated EER (%)	Dimensions
Point-based coif06	35.9	97
Point-based daub04	32.72222	80
Point-based lege04	35.98888	100
Point-based lege06	29.57777	52
Area-based coif06	2.45555	66
Area-based daub04	3.82222	94
Area-based lege04	1.28888	80
Area-based lege06	1.88888	55

This table compares the performance of wavelet feature spaces on the development dataset

derived from the *vertical* GRF component, and 21 were derived from the *medial-lateral* GRF component. This finding was particularly interesting because it contradicted the conclusion in [12] that the GRF vertical component is best suited for subject recognition.

4.5 Summary

This chapter presented the concept known as feature extraction as a means to transform large noisy data sample spaces into smaller useful feature spaces, ideally retaining only the information most relevant to the data analysis objective. In our case, the underlying objective was to extract the features from the footstep GRF best able to discriminate one individual's GRF from another's, and, to achieve this objective, we analyzed and implemented four different feature extraction techniques. In our research, the terms "feature" and "dimension" were used interchangeably while the process of reducing the size of the feature space was often referred to as dimensionality reduction. Furthermore, it was shown that feature spaces can be described as being either heuristically selected or discovered via machine learning techniques. In our research, both methods for establishing feature spaces were presented, with the geometric space being heuristically defined, and the holistic, spectral and wavelet spaces defined via machine learning.

To optimize each feature extractor so that they best conformed to our GRF data, we performed an optimal value search by calculating and comparing the EER for varying configurations. In each case, the EER was calculated using the KNN classifier from Chap. 6 with the value of K set to 5. After running these results for each feature extractor, we found the wavelet feature space to be the most performant followed by the geometric, spectral and lastly the holistic space. In this chapter, we distinguished machine learning feature extractors as being either

Table 4.7 Feature extraction performance comparison

Feature space GRF recognition performance	
Feature space	EER (%)
Optimal geometric	1.33333
Best holistic	2.55555
Best spectral	2.02222
Best wavelet	1.28888

This table compares the best GRF recognition performance achieved across feature spaces extracted using various extraction techniques

supervised or unsupervised, an important distinction when interpreting the results. In this case, the wavelet extractor was shown to be supervised and as such may have benefited from a positive bias in its results due to its underlying exposure to the boundaries between subject samples. Moreover, the geometric feature extractor, as well as the machine learning extractor in direct feature discovery, could also be considered supervised in the sense that it uses a supervised brute force optimization approach to limit the geometric features selected to only those producing the best performance. Conversely, the holistic and spectral feature extractors were developed with no understanding of the underlying subject divisions and thus could be considered unsupervised. Our top results for each of the aforementioned feature spaces are demonstrated in Table 4.7. In the next chapter, we explore the use of dataset normalization as a means to improve upon these results by assisting the feature extractors in selecting the features that better differentiate the GRF subjects. The next chapter places a particular emphasis on finding and exploiting the relationship between stepping speed and GRF curve signature to help assess the second of the two assertions discussed in our objectives.

References

1. Addlesee, Michael D., Alan Jones, Finnbar Livesey, and Ferdinando Samaria. 1997. The ORL active floor [sensor system]. *IEEE Personal Communications* 4(5): 35–41.
2. Bentley, Paul M., and Edward J.T. McDonnell. 1994. Wavelet transformations: An introduction. *Electronics & Communication engineering Journal* 6(4), 175–186.
3. Castellanos, J. Longina, Gómez Susan, and Valia Guerra. 2002. The triangle method for finding the corner of the L-curve. *Applied Numerical Mathematics* 43(4), 359–373.
4. Cattin, Philippe C. 2002. Biometric authentication system using human gait. Ph.D Thesis. Zurich: Swiss Federal Institute of Technology. Switzerland.
5. Christian Scheiblich. JWave. https://code.google.com/p/jwave/.
6. César Souza. 2009. Principal component analysis in C#. http://crsouza.blogspot.ca/2009/09/principal-component-analysis-in-c.html.
7. Derawi, Mohammad Omar, Patrick Bours, and Kjetil Holien. 2010. Improved cycle detection for accelerometer based gait authentication. In *Sixth international conference on intelligent information hiding and multimedia signal processing*, 312–317, Darmstadt.
8. Jolliffe, Ian T. 2002. Introduction. In *Principal component analysis, springer series in statistics*. New York: Springer, chap. 1, 1–9.

9. Jolliffe, Ian T. 2002. The singular value decomposition. In *Principal component analysis, springer series in statistics*. New York: Springer, chap. 3, 44–46.
10. Li, Deqiang, Witold Pedrycz, and Nicolino J. Pizzi. 2005. Fuzzy wavelet packet based feature extraction method and its application to biomedical signal classification. *IEEE Transactions on Biomedical Engineering* 52(6): 1132–1139.
11. Mostayed, Ahmed, Sikyung Kim, Mohammad Mynuddin Gani Mazumder, and Se Jin Park. 2008. Foot step based person identification using histogram similarity and wavelet decomposition. In *International Conference on Information Security and Assurance*, Busan, 307–311.
12. Moustakidis, Serafeim P, John B. Theocharis, and Giannis Giakas. 2008. Subject recognition based on ground reaction force measurements of gait signals. *IEEE Transactions on Systems, Man, and Cybernetics-Part B: Cybernetics* 38(6): 1476–1485.
13. Orr, Robert J. and Gregory D. Abowd. 2000. The smart floor: A mechanism for natural user identification and tracking. In *CHI '00 Conference on Human Factors in Computer Systems*, 275–276, The Hague.
14. Press, William H., Saul A. Teukolsky, William T. Vetterling, and Brian P. Flannery. 1993. 13.4 power spectrum estimation using the FFT. *Numeric Recipes in C.* chap. 13, 549–558, Cambridge University Press.
15. Rodríguez, Rubén Vera, Nicholas W. D. Evans, Richard P. Lewis, Benoit Fauve, and John S. D. Mason. 2007. An experimental study on the feasibility of footsteps as a biometric. In *15th European Signal Processing Conference (EUSIPCO 2007)*, 748–752, Poznan.
16. Rodríguez, Rubén Vera, John S. D. Mason, and Nicholas W. D. Evans. 2008. Footstep recognition for a smart home environment. *International Journal of Smart Home* 2(2) 95–110.
17. Smith, Lindsay I. 2002. *A tutorial on principal component analysis*. USA: Cornell University.
18. Suutala, Jaakko, and Juha Röning. 2008. Methods for person identification on a pressure-sensitive floor: Experiments with multiple classifiers and reject option. *Information Fusion Journal, Special Issue on Applications of Ensemble Methods 9* 9(1) 21–40.
19. Zhang, Feng. 2011. Cross-validation and regression analysis in high dimensional sparse linear models. PhD Thesis. California, Stanford. USA: Stanford University.

Chapter 5
Normalization

Using a feature extraction technique can assist in the discovery of discriminant features and, in datasets containing sources of *intra-subject* sample variability, feature extraction techniques may identify the discriminant features not affected by such variability. However, when it is possible to identify these sources of variability, it may also be possible to use *normalization* to expose the important features that would otherwise be hidden due to differences in the conditions at the time of sample collection. To determine the variance that can be accounted for by normalization, we must find the sample space attributes that both appear consistently across the sample space and correlate to the conditions experienced during data capture. With regards to using footstep GRF for gait recognition, three previous types of inter-subject sample variability have been used for normalization in existing studies: the observed GRF curve *amplitude* [8, 9], the the foot is on the ground during a step) [5] and the *weight* of the studied subject [5].

For the purpose of our research, we have based our recognition model around the assumption that nothing is known about either the subjects or conditions experienced during data collection, leaving only the GRF signature to analyze. In the absence of additional information regarding the conditions experienced during sample collection, normalization could still be accomplished by scaling and/or shifting the GRF force signatures such that they line up according to some standard set of graphical and/or statistical data properties. Alternatively, for the GRF, normalization could also be accomplished by modeling the relationship between step duration and the GRF force curve, then transforming the location of each feature to the location it would be expected to be located at were sample space step durations aligned. The normalization research presented in this chapter distinguishes our work from previous studies, in that we are, to our knowledge, the first to perform an in-depth analysis on the impact of step duration *model-based normalization* on GRF recognition. In the following sections, we begin our analysis by examining the normalization based on simple traditional linear *scaling* and *shifting* operations, and then we introduce two new normalization techniques built around the modeling of the relationship between step duration and the shape of the GRF force curve. To demonstrate the impact that each normalization technique had on the GRF recognition, we normalized our development dataset with every normalizer and passed the results to our best feature extractors, again using the simple KNN classifier produce our recognition results.

© Springer International Publishing Switzerland 2016
J.E. Mason et al., *Machine Learning Techniques for Gait Biometric Recognition*,
DOI 10.1007/978-3-319-29088-1_5

5.1   Scaling and Shifting

The simplest category of normalization involves the application of a single *scale* and/or *shift* operation to transform all samples in a dataset to a chosen common scale. Ideally, after each sample has been transformed to the new scale, all *intra*-subject variability would be removed with only the *inter*-subject variability remaining; this would allow for perfect subject recognition. For instance, if the GRF force signature shape was unique for each subject, but the amplitude of the signatures in the dataset varied with respect to a constant across all samples, then, by scaling the samples such that each was standardized to a *common maximum amplitude*, all *intra*-subject variance would be removed exposing only the remaining *inter*-subject variance. While it is highly unrealistic to expect such an ideal scenario in highly variable data like the GRF force signature, these techniques can still often result in some degree of variance reduction. For the purpose of our research, we examined the impact of five such normalizers on GRF recognition performance: the L^∞ normalizer, the L^1-normalizer, the L^2-normalizer, linear time normalization, and score normalization.

The L^∞-, L^1-, and L^2-normalizers are the most basic of our target normalizers. Using these techniques, normalization is accomplished by *scaling* each GRF force signature in our dataset via the inverse of its L^∞ norm (5.1), L^1-norm (5.2), or L^2-norm (5.3), respectively. Of these three, only the L^∞ normalization technique had previously been used to normalize the footstep GRF for recognition purposes; however, the L^1-and L^2-normalization techniques have been used in a number of image-based recognition studies [6, 14, 15], and, in [3], it was found that, by performing L^1-or L^2-normalization after PCA compression, facial recognition rates increased by as much as 5 %.

Each of the *L-type* normalizers produces a dataset reflecting the characteristics of its given norm. Scaling a dataset using the L^∞ norm (5.1), for a sample of length N with a value at dimension i of y_i, will result in every data sample having a *maximum force amplitude* of 1. Alternatively, scaling a dataset by the L^1-norm (5.2), also known as the *Manhattan norm*, will result in every data sample having a *unit area* of 1. Moreover, scaling by the L^2-norm (5.3), which is also referred to as the *Euclidean norm*, will result in every data sample having a *unit length* of 1.

$$\|y\|_\infty = \max_{i \in N}(|y_i|) \tag{5.1}$$

$$\|y\|_1 = \sum_{i \in N} y_i| \tag{5.2}$$

$$\|y\|_2 = \sqrt{\sum_{i \in N} y_i^2} \tag{5.3}$$

Our implementation of the three aforementioned normalizers used the combined feature space for all eight GRF force signals in the calculation of their norms (i.e., in the case of the L^{∞} normalizer, the maximum amplitude would be the single highest force value recorded in any of a sample's eight GRF signals). In the case of the *optimal geometric* feature space, the optimal features were recalculated using the normalized data and normalization was only applied to the *force* values with no alterations made to statistical or temporal feature values. The results achieved after applying these normalizers to our development dataset and best performing feature extractors are demonstrated in Table 5.1. Both the L^1-and L^2-normalizers led to a significant increase in GRF recognition performance over the equivalent non-normalized results when combined with our best holistic feature extractor; however, all *other* uses of the L-type normalizers demonstrated a large reduction in the recognition performance.

While the L-type normalizers involved aligning data samples according to their amplitude-based norms, data can also be normalized with respect to its temporal properties. One method for performing the temporal normalization, which was previously used for footstep GRF recognition in [5], is *linear time normalization* (LTN). LTN refers to the process of *linearly aligning* the *phase* of data samples with differing durations into a *standard* reference frame such that the samples' proportionate temporal properties can be directly compared. Its performance depends on the assumption that signal duration bears no influence on signal amplitude. Figure 5.1 demonstrates an ideal example of the transform performed by LTN to temporally align two different-duration data samples from the same subject. Under the ideal scenario, performing LTN will result in the phase and amplitude of any *same*-subject samples lining up at a position different from that of any other subject.

Table 5.1 L-type normalizer performance

Feature extractor	Cross-validated EER (%)	Dimensions	EER improvement (%)
L^{∞} *normalized feature extractors*			
Optimal geometric	2.14444	27	−60.8
Best holistic	3.26666	13	−27.8
Best spectral	3.55555	16	−75.8
Best wavelet	2.71111	100	−52.4
L^1-*normalized feature extractors*			
Optimal geometric	1.46666	47	−9.9
Best holistic	2.04444	13	20
Best spectral	2.87777	15	−42.3
Best wavelet	1.63333	97	−26.7
L^2-*normalized feature extractors*			
Optimal geometric	1.48888	69	−11.6
Best holistic	2.05555	16	19.5
Best spectral	2.74444	15	−35.7
Best wavelet	1.56666	76	−21.5

This table demonstrates the change in performance achieved by the L-type normalizers against the best performing feature extractors from Chap. 4.

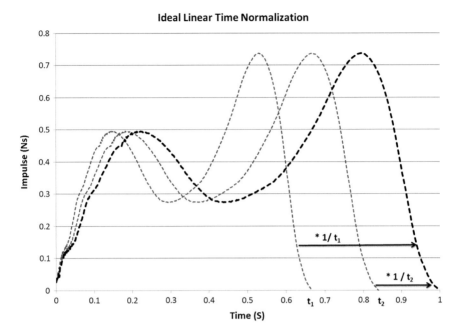

Fig. 5.1 Ideal linear time normalization. This figure demonstrates the *scaling* of the samples ending at t_1 and t_2 by LTN to a common length of 1. Under the ideal scenario shown here the samples line up perfectly when compared in the same *phase*

The method of implementation used to perform LTN depends on the *type* of feature set being normalized. If the features to be normalized have a full or partial temporal component, LTN can be accomplished via scaling them by the *ratio* of a chosen standardized signal duration (i.e., 1 sec) to the duration of the signal being normalized (i.e., 0.71 sec). On the other hand, if the features to be normalized represent the amplitude records of varying length time series recorded at a *standard* sampling rate, then LTN can be accomplished by *resampling* each signal to a standard number of records [2]. In our analysis, this meant using the ratio-based scaling LTN method to normalize the *time* and *area* features in our geometric extractor, and the resampling LTN method to normalize the input to our machine learning-based feature extractors. As the *area-based* sample representation used on the input for all of our best feature extractors already performed resampling in its derivation, to accomplish LTN for our non-geometric feature spaces, we simply took the set of aligned areas produced by the approach and *rescaled* each sample so to remove the time component from each of the representation's area features.

The results of LTN on the applicable normalizers are shown in Table 5.2. LTN was *not applicable* for the *spectral derivative magnitude feature set* since as taking the derivative of the set already negated the time dimension. Looking at our results, the LTN technique achieved a better recognition performance than its non-normalized equivalents for each of the applicable feature extractors. This suggest that, by

Table 5.2 LTN normalizer performance

LTN normalized feature extractors			
Feature extractor	Cross-validated EER (%)	Dimensions	EER improvement (%)
Optimal geometric	1.03333	46	22.5
Best holistic	2.3	16	9.9
Best wavelet	1.1	97	14.6

This table demonstrates the *change* in performance achieved by the LTN normalizer against the best performing feature extractors from Chap. 4

removing the influence of step duration via normalization, we can increase GRF-based gait recognition performance.

An alternative to the previously described scaling normalizers is a category of normalization known as *score normalization*. These normalizers are widely used in statistics and have also found their way into biometric applications such as speech recognition [1]. For the purpose of our research, we applied the standard score (or *Z-score*) normalizer (5.4) to our development dataset. This normalizer shifts and scales the data such that every sample in the dataset will have a *mean* value of zero and *standard deviation* of one; to accomplish this, each i of the N-dimensions (x_i) in a sample has the sample's pre-computed mean (μ) subtracted and is divided by the precomputed standard deviation of the sample (σ). This Z-score normalizer was previously applied to the training samples in the calculation of the PCA *correlation matrix* for our holistic feature extraction technique in Chap. 4; however, in this section the feature extractor is left unaware of the normalization and the Z-score normalizer is applied to all data samples before deriving any feature extraction transformations. The result obtained after applying the Z-score to our development dataset, demonstrated in Table 5.3, showed a slight decrease in recognition performance when the normalizer was applied to the holistic area-based feature extractor and a substantial decrease in recognition performance for all other feature extractors.

$$x_{i \in N} = \frac{x_i - \mu}{\sigma} \tag{5.4}$$

Table 5.3 Z-score normalizer performance

Z-score normalized feature extractors			
Feature extractor	Cross-validated EER (%)	Dimensions	EER improvement (%)
Optimal geometric	2.21111	20	−65.8
Best holistic	2.71111	15	−6
Best spectral	3.4	14	−68.1
Best wavelet	1.7	75	−31.8

This table demonstrates the *change* in performance achieved by the *Z-score* normalizer against the best performing feature extractors from Chap. 4

Looking back at Table 5.1 through Table 5.3, we demonstrated that by performing normalization we could increase the GRF-based gait recognition performance in three of our feature extractors. Yet our results varied greatly and only the LTN normalizer was able to achieve improved performance for each of its applicable feature extractors. It should be noted that these normalizers modeled our dataset in a way that allowed each sample to be compared in a common scale, but none of them were capable of actually modeling the proposed *relationship* between the *step duration* and the *shape* of the eight GRF signal curves. In [12], it was suggested that such a relationship exists and, if this were the case, then by using a normalizer capable of learning and *modeling* this relationship we may be able to significantly improve the GRF recognition performance for our chosen feature extractors. In the next section, we examine this relationship and its application in GRF normalization.

5.2   Regression

To discover and apply the proposed relationship between step duration and the GRF signal curves, we developed a *new* normalization approach based on the derivation of regression models. In this new approach, which we refer to as *localized least squares regression* (LLSR), we derived a set of models able to predict the position that each feature in the dataset would be expected to occupy were the underlying sample step durations aligned. The ability to convincingly and consistently demonstrate an increase in GRF recognition performance using this technique would, together with the results of our LTN in the previous section, support one of the two primary assertions of this book: that a relationship, useful to recognition, exists between the step duration and the GRF force signature.

The LLSR normalization technique draws from an area of data analysis known as *analysis of covariance* (ANCOVA) [7]. Using ANCOVA, the effect of the *covariate*, a variable that has a predictable influence on a data sample being analyzed, is removed from a set of analyzed samples. This is accomplished by aligning each sample according to the *linear relationships* that model the location of sample features with respect to the examined covariate. In our GRF dataset, we treat the *step duration*, the total time recorded from the instant the heel first touches the force plate to the time the toe exits the force plate, as the covariate. In the ideal scenario, as shown in Fig. 5.2, if we knew of a perfect linear relationship between the step duration and the amplitude of a GRF feature, then removing the differences in step duration between the set of samples for any given subjects would result in the complete removal of *intra*-subject variance leaving only the *inter*-subject variance remaining. However, in a practical scenario, given our GRF data, the covariate alignment would simply be expected to result in a *proportionate decrease* of *intra*-subject variance with respect to *inter*-subject variance.

To derive the proposed relationship between the step duration covariate and our data samples, shown in (5.5), we use a regression estimation approach known as

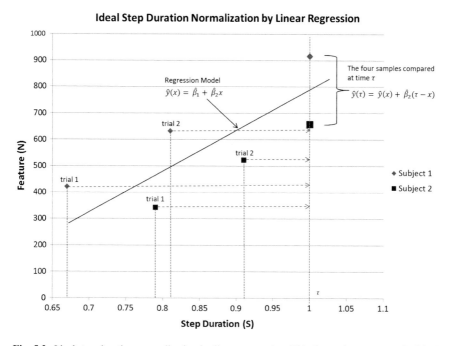

Fig. 5.2 Ideal step duration normalization by linear regression. This figure demonstrates the ideal application of a *linear model* to normalize the features for four samples with respect to *step duration*

least squares regression (5.6). This technique obtains an estimate of the linear relationship between two or more variables by solving for the vector $\boldsymbol{\beta}$ that minimizes some of the *residuals* (or shortest distances) between itself and the locations of the examined features.

Prior to performing regression, to reduce the potential for undesirable data representations being introduced as a result of the inclusion of multiple subjects in our regression analysis, we adjust the sample features such that the sample with the shortest step duration for each subject has its features *shifted* to the coordinate (0, 0). Next, we use the same shifting transform to shift all features in each of the other samples belonging to respective subject so that they end up in the same relative position with respect to the *zero-centered* sample; this can be thought of as a type of *calibration* (see Fig. 5.3). In (5.5), we can see how this calibration fits into the regression relationship through the subtraction of the total step *time* and *force* value corresponding to the minimum total step duration ($\min_{v \in N_i}(t_{iv})$) in the set of trial samples N_i belonging to subject i, setting us up to find the relationship between *total* step duration t_{ij} and the *amplitude* of a given feature $f(t_{ij})$ for a given subject i and trial sample j. The linear estimate of this relationship between the calibrated groupings of $\boldsymbol{y}$ and $\boldsymbol{X}$ in our development dataset is modeled by $\boldsymbol{y} = \boldsymbol{X}\boldsymbol{\beta}$, where β_1 represents the y-intercept and β_2 the linear slope. To solve for the least squares

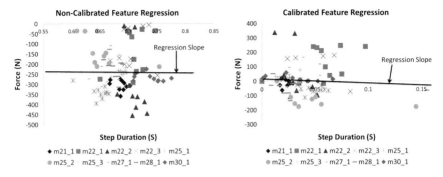

Fig. 5.3 Calibrated versus. non-calibrated feature regression. This figure demonstrates how a relationship between *step duration* and *feature force* becomes more apparent after "calibrating" the data so that the *shortest* step duration for each subject is placed at point (0, 0) and all other feature samples for that same subject are *shifted* around it. In this case, we find the force amplitude for the given feature has a tendency to decrease as step duration increases

approximation of the intercept and slope (represented in the vector $\hat{\beta}$) the linear representation from (5.5) can be rearranged as shown in (5.6).

$$y_{ij} = f\left(t_{ij}\right) - f\left(\min_{v \in N_i}(t_{iv})\right), \; x_{ij} = t_{ij} - \min_{v \in N_i}(t_{iv})$$

$$y = X\beta, \text{ where } y = \begin{bmatrix} y_{11} \\ y_{12} \\ \vdots \\ y_{mn} \end{bmatrix}, \; X = \begin{bmatrix} 1 & x_{11} \\ 1 & x_{12} \\ \vdots & \vdots \\ 1 & x_{mn} \end{bmatrix}, \; \beta \begin{bmatrix} \beta_1 \\ \beta_2 \end{bmatrix} \tag{5.5}$$

$$\hat{\beta} = \left(X^T X\right)^{-1} X^T y \tag{5.6}$$

In our LLSR normalizer, we solve for the linear relationship between each feature and the step duration using *pooled within-group regression* [7]; this approach treats differences in slope (β_2) between different subjects as insignificant and pools the samples from all subjects into the single group for regression analysis. Although pooled within-group regression can lead to the loss of subject specific information, and ideally we would have calculated *individual* regression slopes for each subject, in a typical biometric scenario we would not have enough training samples per subject to generate statistically meaningful results from a subject-specific approach.

Finally, after we have calculated the regression vector for a feature and its covariate, we can then 'remove' the *influence* of step duration on that feature by normalizing all samples containing the feature to the training dataset's *mean step duration*. In this case, demonstrated in (5.7), we can take the relationship derived in (5.5) and (5.6) to model the expected location of a feature $\hat{y}(x)$ for a given step duration x using the y-intercept $\hat{\beta}_1$ and slope $\hat{\beta}_2$. Consequently, if we know the location of a feature and step duration for the sample the feature was recorded in, then we can find the *expected* position of its respective feature in relation to the dataset's mean (μ).

$$\hat{y}(x) = \hat{\beta}_1 + \hat{\beta}_2 x, \quad \hat{y}(\mu) = \hat{\beta}_1 + \hat{\beta}_2 \mu$$

$$\hat{y}(\mu) - \hat{y}(x) = \hat{\beta}_1 + \hat{\beta}_2 \mu - \left(\hat{\beta}_1 + \hat{\beta}_2 x \right) \qquad (5.7)$$

$$\hat{y}(\mu) = \hat{y}(x) + \hat{\beta}_2 (\mu - x)$$

Having so far demonstrated how regression can be used to normalize an individual feature, our LLSR normalizer can be described as the application of this process to *every* feature across our entire dataset feature space. We first calculate the set of regression slopes modeling the relationship between the amplitude of each individual feature and step duration. Then, we use the discovered slopes together with our feature position *modeling function* in (5.7) to derive our *amplitude warping function* (5.8). In this equation, the step duration-based amplitude warping function (Ψ) is demonstrated for an N-dimensional feature set j with a total step duration t_j and an amplitude regression slope A_{2k} for feature k. To complete the formulation, the respective sample feature set S_j belongs to a sample set S with a mean step duration t_μ. Ideally, the feature values s_{jk} would form a perfectly linear relationship for the previously discovered amplitude regression slopes A_{2k} ($\hat{\beta}_2$) and some *subject i*-specific y-intercept $A_{1k}^{(i)}$, in practice, as we can see from Fig. 5.3, subject samples were far from being perfectly linearly aligned. In many cases, however, they did portray a tendency to follow the derived slope, thus our use of the approximation symbol in mapping s_{jk} to its modeled relationship.

$$S_j = \left\{ s_{j1}, s_{j2}, s_{j3} \dots s_{jN} \right\}, s_{jk} \approx A_{1k}^{(i)} + A_{2k} t_j, \ S_j \in S$$

$$\psi(S_j) = \left\{ s_{j1} + A_{21} \left(t_\mu - t_j \right), s_{j2} + A_{22} \left(t_\mu - t_j \right) \dots s_{jN} + A_{2N} \left(t_\mu - t_j \right) \right\} \qquad (5.8)$$

Using (5.8), our step duration-based normalization can finally be accomplished by passing our samples directly into the amplitude warping function. However, for LLSR to be effective each sample in the dataset must have the *same number of features* and these features must be roughly proportional with regards to their positions in *phase* of the GRF curve. When the features are not aligned according to the phase, the regression calculation and amplitude warping function will reflect undesirable phase information. Our geometric features were *heuristically* selected such that the geometric features for each sample lined up according to the phase; but, in the case of our holistic, spectral and wavelet feature spaces, only the *area-based* resampling approaches provided an approximate alignment across samples (samples were roughly aligned by *resampling* to a set of areas prior to extraction). This limitation, however, did not pose a problem when assessing the performance of our LLSR normalizer because our best performing feature spaces all used the standardized area-based resampling approach described in the previous chapter.

The results of running the LLSR normalization with our best performing feature extractors are demonstrated in Table 5.5. Examining these results we can see that

Table 5.4 LLSR-normalized optimal geometric features

LLSR-normalized optimal geometric features			
Feature	Unit	Feature	Unit
$D_{F1Y2MIN2\_F1X1MAX1}$	Force	$D_{F1Z3MIN1\_F1Z3MAX2}$	Force
$D_{F1Y1MIN2\_F1Y2MAX1}$	Force	$D_{F1Y1MAX1\_F1Y2MIN2}$	Time
$D_{F1Z1MAX2\_F1Y2MIN2}$	Time	$D_{F1Z3MIN1\_F1Y1MIN2}$	Time
F1Y1 MIN1	Force	$D_{F1Z1MAX2\_F1Z2MAX1}$	Time
$D_{F1Z2MAX1\_F1Z2MAX2}$	Force	$D_{F1Y1MIN1\_F1X2MAX1}$	Time
$D_{F1X2MAX1\_F1X2MIN1}$	Force	$D_{F1Z2MAX1\_F1X1MAX1}$	Time
$D_{F1Z2MAX1\_F1Z4MAX2}$	Time	$D_{F1Z4MAX1\_F1Z4MIN1}$	Time
$D_{F1Y2MIN2\_F1X2MIN1}$	Force	$D_{F1Z1MAX2\_F1Y1MIN1}$	Time
F1Y1 MIN2	Time	F1Y2 NORM	Force
$D_{F1Z1MAX1\_F1Z1MAX2}$	Force	$D_{F1Y2MIN1\_F1X1MIN1}$	Force
F1Y2 MIN2	Force	F1X1 NORM	Force
$D_{F1Y1MAX1\_F1X1MAX1}$	Force	$D_{F1Z1MIN1\_F1Y2MAX1}$	Time
$D_{F1Z1MIN1\_F1Z3MAX2}$	Time	$D_{F1Z1MAX2\_F1Y1MIN2}$	Time
$D_{F1Z4MIN1\_F1XX1MAX1}$	Time	$D_{F1Z1MAX1\_F1Y2MIN2}$	Time
$D_{F1Z1MAX2\_F1X1MAX1}$	Time	$D_{F1Z2MAX1\_F1Z3MAX2}$	Time
$D_{F1Y1MAX1\_F1Y2MIN2}$	Force	$D_{F1Z1MAX1\_F1Y1MIN2}$	Time
F1Y2 MIN2	Force	$D_{F1Y2MIN2\_F1X2MAX1}$	Force
$D_{F1Z1MAX1\_F1Y2MAX1}$	Time	$D_{F1Y1MAX1\_F1X2MAX1}$	Time
$D_{F1Y1MIN1\_F1X1MIN1}$	Force	$D_{F1Y2MIN1\_F1Y2MIN2}$	Force
$D_{F1Z3MAX2\_F1X1MAX1}$	Time	$D_{F1Z1MAX2\_F1Z3MAX1}$	Time
$D_{F1Y2MIN2\_F1X1MAX1}$	Time	$D_{F1Z1MAX2\_F1Z3MAX2}$	Time
$D_{F1Z2MAX1\_F1Y2MIN1}$	Time	$D_{F1Z3MAX1\_F1Z3MIN1}$	Force
$D_{F1Z1MAX1\_F1Z3MAX2}$	Time	$D_{F1Z2MAX1\_F1Z4MAX1}$	Time
$D_{F1Z3MAX2\_F1Z4MAX2}$	Force	$D_{F1Z2MAX1\_F1Y1MIN1}$	Time
$D_{F1Z4MAX1\_F1Z4MAX2}$	Time	F1X2 NORM	Force
$D_{F1Z1MAX2\_F1Z4MAX2}$	Time	$D_{F1Y1MIN2\_F1X2MAX1}$	Force
$D_{F1Z1MAX2\_F1X2MAX1}$	Force	F1Z1 MAX2	Time
$D_{F1Y2MIN1\_F1Y2MAX1}$	Force		

This table demonstrates the geometric features that were determined to be *optimal* for GRF recognition using the notation presented in the geometric feature extraction section of Chap. 4

the application of the LLSR normalizer to the *optimization* of the geometric feature set produced an 85.1 % reduction in the EER when compared with the non-normalized optimal geometric feature extraction results. Figure 5.4 demonstrates a visualization of a subset of this new feature space formed with the best three LLSR-normalized optimal geometric features, while Table 5.4 demonstrates the complete list of optimal geometric features that make up the space. By comparison, each of our *non*-geometric feature extractors performed poorly when used in conjunction with the LLSR normalizer. It was previously noted that the alignment of features with respect to sample *phase* was implicit in the geometric feature extractor but approximated by a form of resampling in our other feature extractors. If we consider the shape of the GRF phase to be something that can vary between different subjects and even between samples of the same subject (when taken at different walking speeds) then our LLSR regression normalization technique will

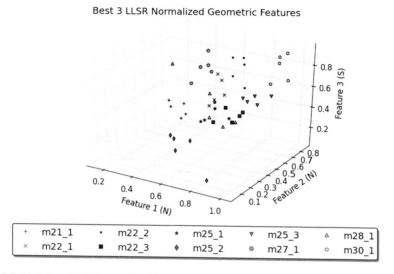

Best 3 LLSR Normalized Geometric Features

+	m21_1	·	m22_2	★	m25_1	▼ m25_3	△ m28_1
×	m22_1	■	m22_3	◆	m25_2	⊙ m27_1	○ m30_1

Fig. 5.4 Best three LLSR-normalized features. This figure demonstrates the *three* features in our geometric feature set that, when combined, result in the best GRF recognition performance. For visualization purposes, we have *scaled* each feature to fall within the range of 0 and 1. When compared with our non-normalized optimal geometric features, these three features demonstrated a 39 % increase in recognition performance

Table 5.5 LLSR-normalizer performance

LLSR-normalized feature extractors

Feature extractor	Cross-validated EER (%)	Dimensions	EER improvement (%)
Optimal geometric	0.17777	55	86.6
Best holistic	2.53333	15	0.8
Best spectral	2.6	16	−28.5
Best wavelet	2.33333	87	−81

This table demonstrates the *change* in performance achieved by the LLSR normalizer against the best performing feature extractors from Chap. 4

end up generating its step duration-to-feature amplitude warping function models using a *distorted* feature alignment, even if all samples were resampled to the same number of dimensions. To address this potential source of error, in the next section, we present an extension of the LLSR normalizer which performs an additional alignment step prior to the generation and application of the regression models.

5.3 Dynamic Time Warping

In the previous section, we found that the LLSR normalization technique was less effective on our non-geometric feature spaces than it was on our geometric feature space. To improve upon the LLSR-normalized GRF recognition performance in our non-geometric feature spaces we adapted the technique to perform a nonlinear sample alignment prior to the generation and application of our regression models. This new technique, which we refer to as *localized least squares regression with dynamic time warping* (LLSRDTW), uses the two-sample *dynamic time warping* (DTW) alignment technique together with the multi-sample *center star* alignment algorithm to accomplish the desired regression alignments. The LLSRDTW normalizer generates two sets of regression models, one reflecting the effect of step duration on the GRF curve *amplitude* and the other reflecting its effect on the GRF curve *phase*. These regression models are generated using a center star aligned training dataset, and, during sample recognition, tested samples are mapped against the center star-aligned training dataset using DTW. The resulting mapping is in turn used to produce phase and amplitude warping functions for the given samples, and the LLSRDTW normalization process is completed by first transforming the samples using the amplitude warping function then passing the amplitude-warped sample through the phase warping function to produce our *normalized* sample.

To understand the LLSRDTW, it is important that we first understand the DTW algorithm that sits at the core of both its warping functions and center star alignment algorithm. The DTW algorithm used in this book takes two samples and derives the *nonlinear* scaling (or path) for each that minimizes the distance between the two. To perform this, we first map every feature in one of the samples to every feature in the other and mark each of the mapped feature pairs with the cost generated by the pair's *global* cost function (5.10). This global cost function is defined recursively, having its actual cost component derived separately via the *local* DTW cost function ($d(i,j)$). The choice of local cost function is highly dependent on the characteristics of the data being observed; for example, when dealing with purely numeric data signals we might choose to use the *absolute difference* local cost function shown in (5.9), which, for the DTW alignment of samples 1 ($f_1(x)$) and 2 ($f_2(x)$), accepts the pair of values (i and j) corresponding to the locally defined sample values at feature index i ($f_1(i)$) and j ($f_2 s j$)) and compares them with the respective *maximum* feature values in each sample. Looking back at the global DTW cost function for the pair of values at index i/j of samples 1 and 2, respectively, it should be noted that it requires an *initialization* value of $i, j = 0$, and the value returned for a given i/j reflects the cost of the best alignment to the pair.

$$d(i,j) = \left| \frac{f_1(i)}{\max(|f_1(x)|)} - \frac{f_2(j)}{\max(|f_2(x)|)} \right| \qquad (5.9)$$

$$D(i,j) = d(i,j) + \min \left\{ \begin{array}{l} D(i-1,j-1) \\ D(i-1,j) \\ D(i,j-1) \end{array} \right\} \qquad (5.10)$$

After mapping an *n*-dimensional sample to an *m*-dimensional sample we would get an $n \times m$ grid like the one shown in Fig. 5.5. In general, a low cost implies that there exists a path to the given pair with strong sample *phase alignment*. To calculate the optimal DTW path, there are four *constraints* that must be taken into

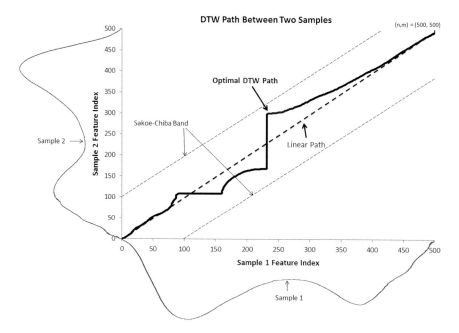

Fig. 5.5 DTW path between two samples. This chart demonstrates how the indices of two samples can form a *grid* of index pairs. It also demonstrates the *optimal* path between the two samples, discovered after calculating the *global* costs for each index pair. This chart also demonstrates the *Sakoe–Chiba Band*, which can be used to constrain the allowable deviation of the optimal path from the linear path

consideration: the path cannot go backwards in time (to smaller feature indices), each feature index must be included in the path at least once, for any given feature position in the DTW path the next position must come from an immediate neighboring feature index pair, and the cost of a feature pair must reflect its local cost as well as its global cost.

In Fig. 5.5, we have demonstrated an *optimal* warping path between an n-dimensional and m-dimensional sample. This path was derived by stepping through the index pairs with the *minimum global cost* from position (0, 0) to position (n, m). The resulting *DTW path* derivation accounts for differences in scale and/or sample length by *repeating* the path indices for one of the samples in regions, where its phase becomes *misaligned* with the other. When the process is finished, the set of DTW feature pairs can be split into a *new* sample space, with both samples approximately aligned according to phase, as shown in the graph on the right in Fig. 5.6. To guide the alignment of our sample GRF signals, we used the *absolute difference* function as our local cost function (5.9), and to improve both the speed and alignment performance of DTW, we used a global path constraint known as the *Sakoe-Chiba Band* [10]. The Sakoe–Chiba Band marks the maximum degree to which the DTW path can deviate from a linear path between the two samples, thus reducing the number of costs that need to be calculated and the potential for

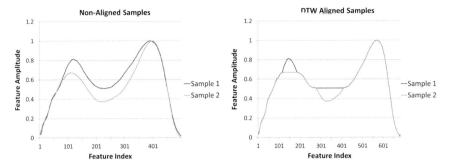

Fig. 5.6 Nonaligned versus. DTW-aligned samples. These graphs compare the nonaligned signals from Fig. 5.5 to those aligned using the derived DTW path. Although the alignment is rigid, it is guaranteed to be *minimum* cost

warping paths that are too distorted to be useful. As a final point, using a technique suggested in a study of DTW for speech recognition by Wang and Gasser [13], our implementation also reduced the potential for path distortions that could arise due to the variation of the sample amplitude by applying the L^{∞} *norm* (sup-norm) during the cost calculation. Consequently, our DTW path is calculated as though the samples used to generate it were aligned according to maximum amplitude when, in fact, the algorithm's output samples do not have their amplitude scaled to any degree.

To this point, we have shown how the DTW algorithm can be used to align any two samples, but to generate our LLSRDTW regression models, we required the alignment of *multiple* samples. Multi-sample alignment can be accomplished with a form of *generalized* DTW [11]; however, such an alignment would incur an *exponential time* complexity penalty making it infeasible for our purpose. Instead, for the purpose of our research, we implemented a *polynomial time* approximation of generalized DTW known as the *center star* algorithm [4]. Using the center star algorithm, if we were given a sample set S of k samples, $S_1, S_2, \ldots S_k$, the first step in the algorithm would be to determine the pairwise alignment costs for every possible combination of samples. In our case, these costs are found by calculating the DTW cost *grid* for every possible sample pair and identifying the pairs according to the final DTW global cost found at the last point of each alignment. A tabulation of these costs shown in Table 5.6 demonstrates how, after calculating these *pairwise* costs, we can expose the single sample S_j with the minimum cost to all others (in the table $S_j = S_2$). For the second step of this algorithm, we create a new aligned sample set S' by first assigning our minimum cost sample (i.e., S_2) to position S_1'. The third and final step of the center star algorithm then involves *iteratively* aligning and including the remaining samples in the new aligned space. For each iteration i from 1 to k, where $i \neq j$, we calculate the DTW alignment DTW (S_1', S_i) to get the two aligned samples S_1'' and S_i'; we then repeat the indices of all samples $S_{r>1}'$ currently in S' at any position for which a S_1' index was repeated in the generation of S_1''; finally, we reassign S_1' to S_1'', add S_i to S', and repeat the

Table 5.6 DTW costs matrix

	S_1	S_2	S_3	S_4	S_5	
S_1	0	4.8226	9.2145	25.791	14.411	$\sum_{i=1...5} D(S_1, S_i) = 54.23973$
S_2		0	17.131	15.362	9.6685	$\sum_{i=1...5} D(S_2, S_i) = \mathbf{51.37646}$
S_3			0	44.824	32.025	$\sum_{i=1...5} D(S_3, S_i) = 118.003$
S_4				0	4.9474	$\sum_{i=1...5} D(S_4, S_i) = 90.92532$
S_5					0	$\sum_{i=1...5} D(S_5, S_i) = 61.05241$

This table demonstrates the tabulation of *global* costs between a set of example samples (S_1 to S_5). The sum of the costs from any given sample to all other samples in the sample set is demonstrated to the right of the table. In this case we can see that the sample S_2 has the *minimum* cost in relation to all others

process for $i \leq k$. The set S' that results from completing this process represents the center star algorithms approximate alignment of S and we refer to the final value for S_1' as the aligned set's center star *template*.

To more thoroughly understand how the center star algorithm works, we can follow the visualization of its application to the alignment of simple biological sequences as shown in Fig. 5.7. In this scenario, rather than repeating indices, sequences are aligned via the inclusion of a space '-' at the positions that produce the minimum cost alignment.

The center star algorithm is a key component in our LLSRDTW normalizer, allowing us to approximately align our training dataset prior to the generation of step duration-to-phase and step duration-to-amplitude regression models. To

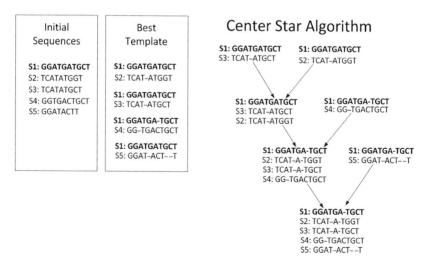

Fig. 5.7 Center star algorithm example. This figure demonstrates how the center star algorithm would typically be used in the alignment of biological sequences. In our sequence alignment, rather than using spaces '-' for alignment, we *repeat* the value at the previous index as previously done in DTW

implement these alignments we determined it would be most effective to apply the center star approximation toward the alignment of each of our 8 GRF *signal* subspaces separately, rather than the across sample as a whole. Consequently, the generation of our regression models over the resulting aligned training sample space produces 16 *phase/amplitude* regression models and 8 center star *templates*, all of which the LLSRDTW normalizer retains to use later in the derivation of its phase and amplitude *warping functions*. An example of the center star alignment for three of our GRF signals is shown in Fig. 5.8, while the corresponding regression models and center star templates are shown in Fig. 5.9.

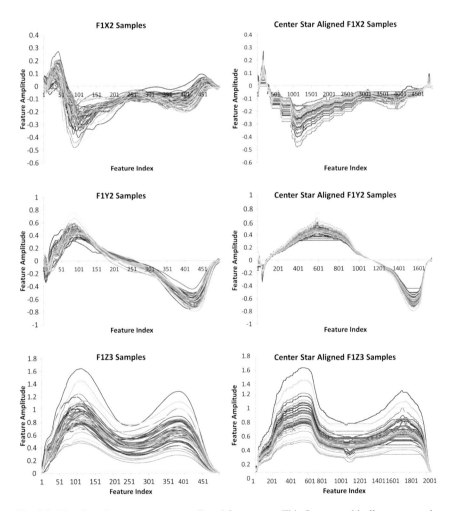

Fig. 5.8 Nonaligned versus. center star-aligned feature sets. This figure graphically compares the center star *aligned* to the *nonaligned* feature sets for three different GRF signals in our training dataset. The expansion of indices that appears in the center star-aligned feature sets is indicative of the *repetition* of feature indices required to perform the alignment

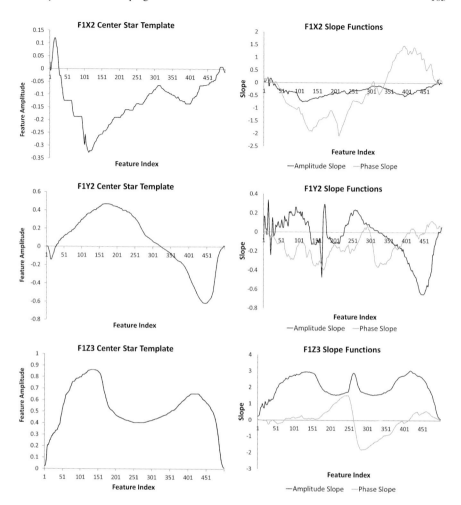

Fig. 5.9 Center star templates and regression models. This figure demonstrates an example of the *best* center star *template* together with its corresponding *amplitude* and *phase* regression model (*slope functions*) corresponding to the three GRF signals alignments demonstrated in Fig. 5.8

$$S_j^c = \{s_{j1}, s_{j2} \ldots s_{jk} \ldots s_{jN}\}, \quad k \in CS(S_j), \quad N = |S_j| \quad S_j^c \in S^c$$
$$s_{jk} \approx A_{1k}^{(i)P} + A_{2k}^P t_j, \quad k \approx \theta_{1k}^{(i)P} + \theta_{2k}^P t_j \tag{5.11}$$

Now that we have seen how the dataset can be aligned we can address the step duration-to-phase regression model. Our LLSRDTW normalizer, unlike our LLSR normalizer, was designed with the ability to determine the *degree* to which the phase of each sample was *misaligned* by comparing the phase distortions required to align them. Following the requirement that all dataset samples have been *re-sampled* to a common length (N), the alignment will, by *repetition* of indices, alter the relative position of each original feature index with respect to the position of the same feature in the other samples. Thus, we can use the center star sample alignment (*CS*) to derive the *Least Squares phase* regression vector θ^p, in addition to the *amplitude* regression vector A^p, which was previously derived for LLSR (5.8). In (5.11) we can see how A^p for the LLSRDTW normalizer was obtained by applying Least Squares regression (5.6) to the set of *aligned* feature values in the center star-aligned sample set S^c, while θ^p was obtained by applying the Least Squares regression to the set of aligned feature indices k in the center star-aligned feature set. In this equation we again identify the step duration for a given sample feature set S_j as t_j, but also present a representation of the center star templates, phase regression vectors, and amplitude regression vectors that have been *resampled* down to the original N-dimensions for computational efficiency.

The normalization of the samples with respect to step duration via the LLSRDTW normalizer is accomplished using both the amplitude and phase warping functions. However, since these functions depend on our regression models and our regression models reflect the center star-aligned feature space, we cannot directly use them in the derivation of our warping functions like was done for the LLSR normalizer (5.8). Instead, to ensure regression models are correctly aligned when included in our warping functions, we perform several *additional* preprocessing steps during the normalization of a sample. First, we find the DTW path between each N-dimensional signal in the sample and its respective P-dimensional center star template, where in our case $P = N$ due to the prior resampling of the center star template. This process is shown in (5.12) with the application of DTW to align a feature set S_j and the respective center star template S_c. The resulting feature set S_j' represents M-dimensional feature set formed by the DTW path and the relation R describes the mapping of indices between the N-dimensional original feature space and the N-dimensional center star template S_c.

$$S_j = \left\{ s_{j1}, s_{j2}, s_{j3}, \ldots s_{jN} \right\}, \quad S_j \in S$$
$$S_j' = \left\{ s_{j1}, \ldots s_{jl}, \ldots s_{jM} \right\}, \quad l \in DTW_{S_c}(S_j), \quad M = \left| S_j' \right| \qquad (5.12)$$
$$R = \left\{ (1,1), \ldots (n,m) \ldots (N,N) \right\} \quad |R| = M$$

Next, we resample our regression models for each signal to the length of the aforementioned DTW paths (i.e., length M). From here, we use the DTW path *mappings* and resampled regression models together with an *averaging* technique to map the regression models back into the original N-dimensional sample space. This process is demonstrated in (5.13) with the P-dimensional step duration-based *amplitude* and *phase* slope functions (A_2^P and θ_2^P) being resampled to M-dimensions and then aligned to the N-dimensional feature set S_j from (5.12). Again note that in our implementation we *resample* the center star template, amplitude slope function and phase slope function to the dimensionality of the *original* dataset after generating our regression models, thus $P = N$. Furthermore, we use the DTW relationship function R in our formulation for aligning the amplitude and phase slope functions to N-dimensional representations ($A_2^{N'}$ and $\theta_2^{N'}$). In this case, repeated indices in the DTW path are *collapsed* using averaging, so, for example, if we had the indices (1, 1) and (1, 2) in our DTW path then we would find $R(1) = \{1, 2\}$ and $\overline{A_{2R(1)}^{M}} = \left(A_{21}^{M} + A_{22}^{M}\right)/2$.

$$A_2^M = \left\{A_{21}^P, A_{22}^P, A_{23}^P \ldots A_{2M}^P\right\}, \cdots \theta_2^M = \left\{\theta_{21}^P, \theta_{22}^P, \theta_{23}^P \ldots \theta_{2M}^P\right\}$$
$$A_2^{N'} = \left\{\overline{A_{2R(1)}^M}, \overline{A_{2R(2)}^M}, \overline{A_{2R(3)}^M}, \ldots \overline{A_{2R(N)}^M}\right\}, \qquad (5.13)$$
$$\theta_2^{N'} = \left\{\overline{\theta_{2R(1)}^M}, \overline{\theta_{2R(2)}^M}, \overline{\theta_{2R(3)}^M}, \ldots \overline{\theta_{2R(N)}^M}\right\}$$

Having *mapped* the regression models into the original sample space, the LLSRDTW *amplitude warping function* for any given signal in the sample will be almost identical to the simple warping function used by the LLSR normalizer. This step duration-based amplitude warping function (ψ') is demonstrated in (5.14), for a feature set j with a total step duration t_j, belonging to the sample set S, and with a mean step duration t_μ. In this equation, we can see how amplitude warping function was derived by modifying (5.8) to use the N-dimensional amplitude regression slope derived in (5.13).

$$\psi'(S_j) = \left\{\begin{array}{c} s_{j1} + A_{21}^{N'}(t_\mu - t_j), s_{jk} + A_{22}^{N'}(t_\mu - t_j), \ldots \\ s_{jN} + A_{2N}^{N'}(t_\mu - t_j) \end{array}\right\} \qquad (5.14)$$

The LLSRDTW *phase warping function* is slightly more complicated than the amplitude warping function. It first uses the *resampled* regression models to determine the *degree* to which the phase of the feature set must be *warped* to align it with the mean step duration (δ), next it applies L^1-*normalization* to the result (γ), and finally it shifts the L^1-normalized result such that the final result acquires a mean value of 1 (ϕ). These steps ensure the phase warping function only affects the *shape* of the curve but does not in any way alter its amplitude. The formulation for the phase warping function (ϕ) is demonstrated in (5.15) for feature set j with a total step duration of t_j, belonging to the sample set S, with a mean step duration t_μ. The phase warping function should *always* be applied *after* the amplitude warping function, thus in our application the LLSRDTW normalizer can be described as the application of $\phi(\psi(S_j))$; performing the warping functions in reverse order would lead to the amplitude warping function being computed over a *different* phase from that in which it was derived, thereby producing a misaligned amplitude regression.

$$\delta\left(S_{j}\right) = \left\{\theta_{21}^{N'}\left(t_{\mu} - t_{j}\right), \theta_{22}^{N'}\left(t_{\mu} - t_{j}\right) \dots \theta_{2N}^{N'}\left(t_{\mu} - t_{j}\right)\right\}$$

$$\gamma\left(S_{j}\right) = \frac{\delta\left(S_{j}\right)}{\delta\left(S_{j}\right)_{1}} \qquad\qquad (5.15)$$

$$\phi\left(S_{j}\right) = S_{j} \times \left(\gamma\left(S_{j}\right) - \overline{\gamma\left(S_{j}\right)} + 1\right)$$

As mentioned earlier, in our implementation, LLSRDTW normalization is performed on a given sample by first passing each of the sample's GRF signals through the amplitude warping function, then passing the resulting amplitude-warped feature sets into the phase warping function. To improve the speed and performance of our LLSRDTW normalizer, we tested its underlying DTW algorithm with a set of *Sakoe–Chiba Band* values from *one* to *twenty percent* of the DTW cost grid size and discovered optimal *bandwidths* of 5, 1, and 10 % for our best holistic, spectral and wavelet feature extractors, respectively. The results from the application of our LLSRDTW to our best *non-geometric* feature sets are demonstrated in Table 5.7. Analyzing these results, we discovered that by better aligning our features we were able to achieve a modest increase in performance in all our feature extractors over the *non-aligned* LLSR normalization technique. However, only the LLSRDTW-normalized holistic and spectral feature spaces produced an increase in recognition performance over their non-normalized equivalents. The most notable outcome from performing LLSRDTW normalization was the performance increase achieved when it was applied to the spectral feature space. This technique achieved an increase in recognition performance where all other normalization techniques failed to increase performance. Consequently, our results suggest that by aligning each sample as though they were all taken with the same step duration we can improve GRF recognition performance in both heuristic and machine learning-based feature extractors, thus demonstrating the potential utility of the relationship between the step duration and the shape of the GRF curve.

Table 5.7 LLSRDTW performance

LLSRDTW-normalized feature extractors				
Feature extractor	Cross-validated EER (%)	Band (%)	Dimensions	EER improvement (%)
Best holistic	2.3	5	13	9.9
Best spectral	1.84444	1	17	8.7
Best wavelet	1.45555	10	90	−12.9

This table demonstrates the *change* in performance achieved by the LLSRDTW normalizer against the best performing feature extractors from Chap. 4

5.4 Summary

This chapter demonstrated how normalization techniques could be used to improve the selection of features for footstep GRF-based person recognition. In this chapter, we discussed how variation in intra-subject GRF sample scale could potentially lead to feature extractors missing important features on account of their inability to distinguish differences due to variations in scale from those due to distinctive inter-subject characteristics. To address this issue we reexamined our best feature extractors from the previous chapter, but this time with various normalization techniques applied prior to feature extraction. The normalizers examined used various methods to transform our footstep samples to a common scale in terms of both step duration and amplitude, including well-known uniform scaling and shifting operations and two new dynamic techniques developed for the purpose of this research.

The two new normalization techniques introduced in this chapter (LLSR and LLSRDTW) were created to test our assertion that a potentially useful relationship exists between stepping speed (or step duration) and the GRF force signature. The LLSR normalizer attempted to model this relationship via the derivation of a series of individual regression functions, while the LLSRDTW was designed to improve upon the LLSR in machine learning-based feature extractors using the technique known as DTW to align key sample data points prior to deriving the regression models. The best results obtained after applying these and the scaling and shifting normalizers to our top feature spaces in the development dataset are shown in Table 5.8 (again we used the KNN classifier to acquire these results). Although the LLSRDTW normalizer was found to result in a clear improvement over the LLSR normalizer in all applicable feature spaces, it only proved to be the most performant normalizer in the spectral feature space. On a whole, normalization was shown to improve GRF recognition in all feature spaces, but with regards to the best normalizer our results were divided with no two feature spaces achieving their best results over a shared normalizer. However, it must be noted that we were able to improve the recognition performance over each feature space when using a normalizer that accounted for step duration, which would appear to support our assertion regarding the relationship between stepping speed and the GRF force signature.

Table 5.8 Normalizer performance cross comparison

Normalizer GRF recognition performance		
Feature space	Best normalizer	EER (%)
Optimal geometric	LLSR	0.17777
Best holistic	L1	2.04444
Best spectral	LLSRDTW	1.84444
Best wavelet	LTN	1.1

This table compares the best GRF recognition performance achieved across each feature space when used in combination with a normalization technique

In this chapter and the one preceding it, we have examined two parts of the biometric system that could jointly be described as data preprocessing. The application of these techniques makes it easier to identify important characteristics but they do not have the ability to distinguish GRF subjects on their own. In the next chapter, we explore the biometric system component that is responsible for learning the patterns within the preprocessed data and using them to perform footstep GRF-based recognition.

References

1. Barras, Claude, and Jean-Luc Gauvain. 2003. Feature and score normalization for speaker of cellular data. *IEEE International Conference on Acoustics, Speech, and Signal Processing (ICASSP '03)* 2(2) II—49–52. Orsay.
2. Byrd, Dani, Sungbok Lee, and Rebeka Compos-Astorkiza. 2008. Phrase boundary effects on the temporal kinematics of sequential tongue tip consonants. *Journal of the Acoustical Society of America* 123(6): 4456–4465.
3. Cao, Zhimin, Qi Yin, Xiaoou Tang, and Jian Sun. 2010. Face recognition with learning-based descriptor. In *IEEE Conference on Computer Vision and Pattern Recognition*, San Francisco. 2707–2714.
4. Gusfield, Dan 1993. Efficient methods for multiple sequence alignment with guaranteed error bounds. *Bulletin of Mathematical Biology* 55(1): 141–154.
5. Moustakidis, Serafeim P., John B. Theocharis, and Giannis Giakas. 2008. Subject recognition based on ground reaction force measurements of gait signals. *IEEE Transactions on Systems, Man, and Cybernetics-Part B: Cybernetics* 38(6): 1476–1485.
6. Nistér, David, and Henrik Stewénius. 2006. Scalable recognition with a vocabulary tree. In *IEEE Computer Society Conference on Computer Vision and Pattern Recognition (CVPR'06)*, New York. 2161–2168.
7. Quinn, Gerry P., and Michael J. Keough. 2002. Analysis of Covariance. In 1 (Ed.), *Experimental design and data analysis for biologists*. Cambridge University Press. 339–358.
8. Rodriguez, Rubén Vera, Nicholas W.D. Evans, Richard P. Lewis, Benoit Fauve, and John S. D. Mason. 2007. An experimental study on the feasibility of footsteps as a biometric. In *15th european signal processing conference (EUSIPCO 2007)*. 748–752, Poznan.
9. Rodriguez, Rubén Vera, John S.D. Mason, and Nicholas W.D. Evans. 2008. Footstep recognition for a smart home environment. *International Journal of Smart Home* 2(2): 95–110.
10. Sakoe, Hiroaki, and Seibi Chiba. 1978. Dynamic programming algorithm optimization for spoken word recognition. *IEEE Transactions on Acoustics, Speech, and Signal Processing* 26 (1): 43–49.
11. Sanguansat, Parinya. 2012. Multiple multidimensional sequence alignment using generalized dynamic time warping. *WSEAS Transactions on Mathematics* 11(8): 668–678.
12. Taylor, Amanda J., Hylton B. Menz, and Anne-Maree Keenan. 2004. The influence of walking speed on plantar pressure measurements using the two-step gait initiation protocol. *The Foot* 14(1): 49–55.
13. Wang, Kongming, and Theo Gasser. 1997. Alignment of curves by dynamic time warping. *The Annals of Statistics* 25(3): 1251–1276.
14. Wang, Xingxing, Li Min Wang, and Qiao Yu. 2012. A comparative study of encoding, pooling, and normalization methods for action recognition. *ACCV'12 Proceedings of the 11th Asian conference on Computer Vision Volume Part III*. 572–585, Daejeon.
15. Zhu, Qiang, Shai Avidan, Mei-Chen Yeh and Kwang-Ting Cheng. 2006. Fast human detection using a cascade of histograms of oriented gradients. *IEEE Computer Society Conference on Computer Vision and Pattern Recognition*. 1491–1498, New York.

Chapter 6
Classification

The final step in performing biometric recognition after preprocessing the data with normalization and feature extraction involves the application of a *classifier* to perform recognition over the transformed sample space. Although classification could theoretically be applied directly to the sample set with no prior preprocessing, we opted against doing so, because, in addition to exposing features to the classifier that might otherwise be missed by *overfitting*, the use of our preprocessors also reduced the dimensionality of the samples' feature space to a size that could be effectively handled by all popular classifiers. Without needing to take computational efficiency into account, the goal of the classifier then became to find the *boundaries* in the transformed sample space that best separate the feature spaces of the different subjects. Internally, classifiers use a number of tricks to discover these boundaries; however, classifiers are also subject to *initialization parameters* that can be adjusted to improve recognition performance. Knowing this, our classification goal, for any given classifier, can be addressed by solving an optimization problem; namely, discovering the classification parameters that optimize recognition performance.

In our discussion in Chap. 3 we described how classifiers can be categorized as being based on either *generative* or *discriminative* models. We could also further distinguish classifiers as being either *instance*-based (lazy learning-based) or *eager learning*-based [31]. In *instance*-based classifiers the training phase involves storing all training samples to memory and all calculations needed for recognition are performed at the time of recognition. Conversely, in *eager learning*-based classifiers the training phase involves the derivation of a function that is able to transform any given subject/sample pair into a representation of a decision reflecting whether the given subject corresponds to the given sample. In the following sections we examine classifiers covering each of the two classifier models and learning strategies. For each classifier we discuss the *parameter optimizations* that were used to improve recognition performance, while, as a final assessment of our various GRF recognition strategies we cross compare the full set of classifiers with our best performing feature extraction and normalization strategies from Chaps. 4 and 5.

© Springer International Publishing Switzerland 2016
J.E. Mason et al., *Machine Learning Techniques for Gait Biometric Recognition*,
DOI 10.1007/978-3-319-29088-1_6

6.1   *K*-Nearest Neighbors

The most common category of classifier applied by previous GRF recognition studies [5, 26, 27, 29, 33, 41] was the *K-Nearest Neighbors* (KNN) [43] classification technique. This was the classification technique used in the two preceding chapters to assist with the optimization of our feature extractors and normalizers. The KNN classifier is an *instance*-based *discriminative* classification technique that uses the distances between a given sample and its *K-nearest* training samples to determine the sample's identity. The decision is based on a voting scheme under which the subject corresponding to the *majority* of the *K*-nearest neighbors is identified as the owner of the sample. This technique comes in many variants and can be altered to perform *verification* rather than identification by simply taking the expected identity as a verification parameter then accepting or rejecting the verification requests based on the discovery of a matching identity in the majority of the *K*-nearest neighbors.

The KNN recognition algorithm in its most basic form is relatively straightforward. During its *training* phase the classifier simply stores all training samples to *memory*. Subsequently, when performing recognition on a tested sample the KNN classifier begins by calculating the *Euclidean distance* (6.1) between the sample and every sample in the training set; with the vectors p and q representing the features sets for the two different samples.

$$d(\boldsymbol{p}, \boldsymbol{q}) = \sqrt{(q_1 - p_1)^2 + (q_2 - p_2)^2 + \cdots + (q_n - p_n)^2}$$
$$= \sqrt{\sum_{i=1}^{n}(q_i - p_i)^2} \qquad (6.1)$$

Next, the training samples are ordered according to their distance from the tested sample and the samples with the K smallest distances are used to represent a *voting set*. Finally, the identity assigned to the sample would correspond with the owner of the *majority* of samples in the voting set, or, in the event of a draw, the owner of the samples in the voting set with the *shortest distance* to the tested sample. If this identity matches the identity provided for recognition then the algorithm would accept it as a match, otherwise it would be rejected. A simple visualization of the KNN identification process for a 2-dimensional feature space is demonstrated in Fig. 6.1. Using this simple identification scheme, the KNN classifier can be observed as having a key advantage over *non-instance*-based classifiers in that it does not need to be retrained when *new* samples are *added-to* or *existing* samples *removed-from* its training set.

In our experimental design, we established that the verification EER would be used to assess GRF recognition performance, and that this value could be derived by first configuring our classifier to return *posterior probabilities* then scaling the classifier's *acceptance threshold* to the point at which the FAR and FRR intersect.

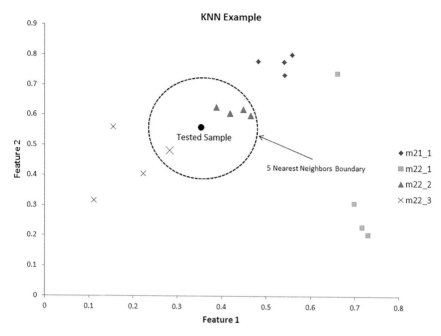

Fig. 6.1 KNN example. This graph demonstrates the samples that would form the *voting set* in a 2-dimensional feature space with the KNN *K* value set to 5. In this case only the subject *m22_2* would be accepted as a recognition match for the given test sample

The basic variant of the KNN classifier described in the previous paragraph does not return a probability, but, rather, a *true* or *false* decision based directly on the results of the voting scheme. However, in [41] Suutala and Röning demonstrated that posterior probabilities, which reflect the *likelihood* of a given subject matching a given sample, could be estimated by *counting* the occurrences of the given subject in the *K*-nearest training samples for the given sample (6.2). This *unweighted* estimation of the posterior probability is acquired by finding the proportion of the number of occurrences of subject *i* (measured as k_i) to the number of selected *k*-nearest neighbors for a tested sample *x*, with w_i being the identifier for subject *i* and $P(w_i|x)$ the estimate of the *probability density function*.

$$P(w_i|x) = \frac{p(x|w_i)p(w_i)}{\sum_{j=1}^{c} p(x|w_j)p(w_j)} \approx \frac{k_i}{k} \tag{6.2}$$

Unfortunately, the KNN posterior probability generation technique used in [41] can be subject to undesirable *biases* from *outlier points* when either the number of training samples per subject is limited or the value chosen for *K* is small. To address this problem we used a form of *weighted* KNN [43], for which the *K* samples in the voting scheme were given weights *inversely proportional* to their *distances* from the sample being recognized. Using this weighted KNN technique, we were then able to alter (6.2) to reflect the assigned weight values, as shown in (6.3), and thus acquire posterior probabilities less affected by *small sample size*-induced biases. In this weighted equation the occurrences of subject i are weighted according to the distance between the occurrence u_{k_j} and the tested sample x. Only samples belonging to the tested subject i are counted in the numerator, as demonstrated through the values returned by the expression $v_{k_j} = i$.

$$P(w_i|x) \approx \frac{\sum_{j=1}^{k} \frac{\left(v_{k_j}=i\right)}{d\left(u_{k_j},x\right)}}{\sum_{j=1}^{k} \frac{1}{d\left(u_{k_j},x\right)}}, \quad \left(v_{k_j} = i\right) = 1, \ \left(v_{k_j} \neq i\right) = 0 \qquad (6.3)$$

There are two configurable *inputs* that must be accounted for when implementing the KNN classifier: the parameter K used to represent the *size* of the voting set, and the *sample feature sets* used in the training and testing of the algorithm. Examining our sample inputs we noticed that each of the features in the sample feature space was represented over a different *scale* reflecting the *degree* of *absolute variance* in the part of the feature space from which the feature was derived. Consequently, features represented in larger scales would gain a large influence on the classification result regardless of their discriminative ability. Ideally, each feature would be scaled in proportion to its discriminative ability. However, determining the correct proportions for such scaling is generally computationally infeasible, so instead we decided to use the approach taken in [41] and gave each feature *equal* influence in the recognition decision. We accomplished this by finding the *minimum* and *maximum* values for each feature in our training dataset, setting these values to 0 and 1, and *rescaling* all training and testing sample features with respect to these feature space boundaries. In addition to this input rescaling we attempted to optimize the value of K to improve our recognition performance. To do so we used a *brute force* approach, comparing the *cross-validated* EER results returned for 10 different values of K across each of our best performing preprocessor configurations, as demonstrated in Fig. 6.2.

As shown in Fig. 6.2, the KNN classifier tended to perform worse when the K value was greater than the number of samples used for training each subject. We also observed a tendency for our best results to occur when the K value was set equal to the number of *per-subject* training samples with $K = 5$, though it must be noted that because our preprocessor optimization was done using a K value of 5 our results were *positively biased* toward this value of K. Yet, despite this bias we were still able to achieve better results in half of our preprocessors using other values of K, as shown in Table 6.1. Comparing these findings with those in [33], the only previous GRF recognition study to identify an optimal K value, we found that *all but one* of our K values represented a larger *percentage* of the number of samples used for training a single subject than was found by Rodríguez et al. Yet, the KNN classifier used in [33] was *not weighted*, and, of the previous GRF recognition

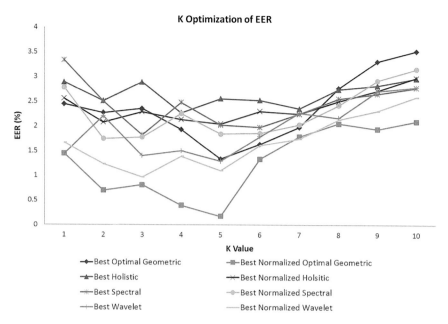

Fig. 6.2 KNN parameter optimization. This figure demonstrates the cross-validated EER achieved by our best feature extractors and normalizers for the values of *K* from 1 to 10

Table 6.1 KNN classifier performance

Optimal KNN classifier results				
Preprocessor	K	Thres.	CV EER (%)	EER incr. (%)
Best optimal geometric	5	0.3359	1.33333	0
Best normalized optimal geometric	5	0.3921	0.17777	0
Best holistic	4	0.2765	2.26666	11.3
Best normalized holistic	5	0.3015	2.04444	0
Best spectral	3	0.2515	1.82222	9.8
Best normalized spectral	2	0.2859	1.74444	5.4
Best wavelet	5	0.1703	1.28888	0
Best normalized wavelet	3	0.2921	0.96666	12.1

This table demonstrates the best performance achieved by the KNN classifier for each preprocessing technique. The *threshold* shown is the threshold at which the EER value was calculated (a value between 0 and 1 derived from the *raw posterior probability output*) and the EER improvement represents *improvement* in recognition performance achieved by the *K* value optimization over the results calculated in the previous two chapters

studies, only [27] involved the use of a weighted KNN algorithm similar to the one used in this section (Fuzzy KNN [18]). Thus our KNN-specific findings could not be directly compared with any previous GRF studies.

6.2   Multilayer Perceptron Neural Network

In many classification problems the distributions of the classes being classified do not form the closely bundled *symmetric clusters* optimal for classification when using distance-based classifiers like the KNN. Instead the *boundaries* that separate the various classes (in our case subjects) may be *irregular* with sharp cut-off points as shown in Fig. 6.3. These boundaries can be difficult to define geometrically; however, they can usually be approximated to a high degree of accuracy using a *machine learning* structure known as an *Artificial Neural Network* (or Neural Network). Neural Networks derive from early attempts to model the problem-solving abilities of the connected neurons in the human brain [36], and have proven particularly useful in the field of *pattern recognition* [21]. These networks are typically represented as a *directed graph* consisting of a set of interconnected processing units, or *nodes* (Fig. 6.4), with one subset of the nodes accepting input data and another output subset presenting the results of any processing that was performed as the input data passed through the network. The structure represented by the arrangement of these nodes can be categorized as being either *feed-forward* or *recurrent*. In *feed-forward* networks, information passes between the input nodes and the output nodes without ever passing through the

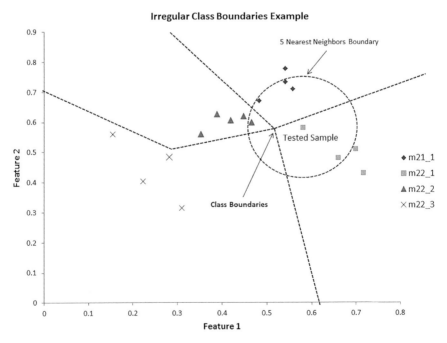

Fig. 6.3 Irregular class boundaries example. This figure demonstrates a sample "Tested Sample" that would be attributed to the wrong class by the KNN classifier, but would be correctly matched were the identified class boundaries used during classification

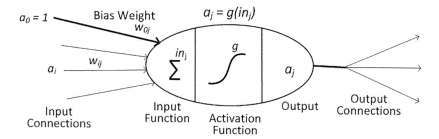

Fig. 6.4 Artificial neural network node. This figure demonstrates an individual ANN node. The node accepts a series of weighted inputs including a bias input, which first increases or lowers the net input where needed [11], transforms it using the activation function (*g*), and returns the transformed output to any output connections

same node twice, while in *recurrent* networks, processed output will end up getting passed back into nodes that have already seen the original input, allowing the network to become aware of *state*. For the purpose of our research, we based our GRF Neural Network analysis on a popular forward-feed network structure known as a *Multilayer Perceptron* (MLP) [32], a structure which previously achieved strong GRF recognition performance in [41].

The MLP classifier is an *eager learning*-based classification approach belonging to the *discriminative* category of classifiers. At its core is a *feed-forward* Neural Network with nodes arranged into three or more *layers* including an *input* layer containing all input nodes, an *output* layer containing all output nodes, and one or more *hidden* layers, which sit between the input and output layers. Within each layer every node is *fully connected* with all nodes in the neighboring layers, but no connections are made between the nodes of a single layer or to nodes in non-neighboring layers. All connections between nodes contain the *product* of output returned by the node feeding into the connection and a *weight* value, which is determined during the training of the network. Furthermore, aside from input nodes, which allow data input to pass directly through them, all other nodes in the MLP network *sum* and pass their input data through a *nonlinear activation function*. For our research, we have chosen to assess the MLP architecture demonstrated in Fig. 6.5. This architecture, which reflects the one used in [41], contains three layers and uses the *logistic sigmoid function* (6.4) as its activation function. Each node in the input layer corresponds with an individual feature from the feature space used to train the MLP, while each output node corresponds with a different subject from the training data. With this structure in place, the *output* for a particular subject node will be a value between 0 and 1, signifying the *confidence* for which the network has determined any given input features match those previously learned for the given subject.

$$g(x) = \frac{1}{1 + e^{-x}} \qquad (6.4)$$

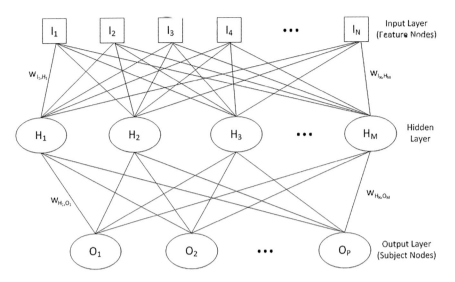

Fig. 6.5 Three layer mlp architecture. This diagram demonstrates our chosen 3 layer MLP architecture. The nodes in the input layer contain no activation function and pass any received data directly into the connection with the hidden layer, while the hidden and output nodes were configured to use a logistic sigmoid activation function. Each node is connected with every node in its immediate neighbors and all connections apply a weight to the data going into the following layer, as demonstrated on the edge connections

The MLP architecture described to this point gives us the means to learn how to identify subjects based on their features, but before it can perform any classifications it must first undergo a *training process*. As previously mentioned, the data passed between connected nodes in the MLP is multiplied by *weight* values. These weight values can be adjusted so to *minimize* the error rate produced by the classifier for a set of training input–output pairs via a process known as *backpropagation* [8]. The backpropagation process is premised on the fact that the MLP network can be represented as a *vector function* $h_w(x)$ parameterized by its weights (w). So, if we had a pair of input features (x) and an output subject set (y) with a value of 1 corresponding to subject nodes that match the input and 0 for those that do not, then the vector of errors for our network's output nodes would be represented as $E = y - h_w(x)$. Using this set of error values, it is possible to derive the network's *node deltas*, which, for any given connection from node i to node j, would represent the effect of a change in the input to node j on the output of node i [35]. At the output layer these node deltas can be calculated *directly*, while in the input and hidden layers they are computed by *backpropagating* the values of the deltas computed in the deeper neighboring layer, as demonstrated in (6.5). In this equation the term in_i represents the sum of the weighted outputs from node i to node j, which are to be passed back to the *derivative* of i's activation function. The node *deltas* themselves for each node i are identified as δ_i, w_{ij} represents the *weight* between nodes i and j, and E_i the portion of the *error* produced by a given output node i.

$$E = y - h_w(x)$$

$$in_i = \sum_j w_{ij} a_j + \text{bias}_i$$

$$\delta_i = \begin{cases} E_i \times g'(in_i), & \text{output nodes} \\ g'(in_i) \times \sum_k \delta_k \times w_{ki}, & \text{input/hidden nodes} \end{cases}$$

(6.5)

The aforementioned *node deltas* can also be used in conjunction with the computed output from the node feeding into the delta connection to derive the partial derivative *(gradient)* for each of the network's *weights* with respect to the *error* value. This gradient of the error with respect to the weight is shown in (6.6), between nodes *i* and *j*, at the point for which the output of node *j*'s activation function is a_j. A consequence of this relationship is that it means the error value can be *minimized* via the *gradient descent* method, which follows the contour of the error surface in the direction of steepest descent [35].

$$\frac{\partial E}{\partial w_{ij}} = \delta_i \times a_j \tag{6.6}$$

The use of gradient descent in the backpropagation process involves *iteratively* adjusting the network weights via the weight update function shown in (6.7). This process will continue for a predefined *number of iterations* or until the returned error falls below some *maximum threshold*, at which point the network error term can be assumed to have been brought to a *local minimum*. In our chosen weight update function, the speed at which the error is minimized is subject to two constants, namely the *learning rate* ($\in$) and *momentum* (α). The *learning rate* can be used to adjust the *size* of the step taken down the gradient for each weight update, while the *momentum* term prevents the weight updates from *oscillating* between opposing sides of a trough in the error contour, by taking into account the trajectory of the previous iteration [35].

$$\Delta w_{ij}(n+1) = \epsilon\big(\delta_i \times a_j\big) + \alpha \Delta w_{ij}(n) \tag{6.7}$$

Additionally, the *weight update function* can be altered so that it minimizes the error across multiple training samples during a single update by *batching* (summing) the gradients produced by each individual training sample (6.8). This variant of the weight update function allows the process to *simultaneously* reduce the error for all input–output pairs *p* provided at the time of training. It is this batching-based weight update function that represents the final piece of the process needed to train our MLP classifier with our multiple GRF feature–subject pairs.

$$\Delta w_{ij}(n+1) = \epsilon \sum_p \big(\delta_{ip} \times a_{jp}\big) + \alpha \Delta w_{ij}(n) \tag{6.8}$$

Our implementation of the MLP was constructed using the *Encog* Machine Learning Framework for C# [14]. Following the typical convention for MLP weight initialization, we *randomized* our weights prior to performing backpropagation [8], making use of a *seeded* randomizer with a standard seeding to ensure consistency between training runs. We also configured our training process to terminate after either completing 100,000 iterations or when the network's computed error term fell below a value of 0.0000001. This left *four configurable inputs* to be accounted for: the *input features*, the *learning rate*, the *momentum*, and the number of *hidden nodes*. As discussed in the previous section, undesirable *bias* in input features can be mitigated by *rescaling* each feature space to a common scale. For the KNN classifier this was accomplished by rescaling the input features to fall within the range [0, 1]. In our MLP classifier we instead went with a rescaling *range* of [−1, 1], as our chosen activation function, the *logistic sigmoid function*, expects input to be distributed around the *zero* mark. The use of the sigmoid activation function in our network also meant the results produced at our output nodes came out as values between 0 and 1, representing the network's confidence as to whether or not any given input features matched the subject corresponding to any given output node. Thus, unlike in the KNN classifier, we were able to use the output *directly* in our EER calculations, rather than first needing to find posterior probabilities. Having established a means to calculate our EER values, the three remaining configurable inputs were optimized via *exhaustive search*, whereby we tested the system with every combination of 10 different learning rate and momentum terms (the 10 evenly spaced points from 0.1 to 1.0) and 7 different hidden field sizes (the 7 evenly spaced nearest integers from 20 to 140 % of input feature space size), that is, in summary, 700 different combinations of parameters were tested for each of our best performing preprocessors. The best achieved GRF recognition results for each tested configuration of learning rate, momentum, and hidden nodes are demonstrated in Fig. 6.6.

Looking back at the optimization results in Fig. 6.6, we found our best GRF recognition performance was achieved with MLP *learning rate* terms less than 0.5, *momentum terms* less than 0.8, and with the *hidden nodes* numbering over 60 % the size of input feature nodes. The best of these results, broken down by feature extractor, are demonstrated in Table 6.2. In contrast with the findings of [41], in which the MLP classifier led to a 9.4 % improvement in GRF geometric feature space recognition over the KNN classifier, we noticed a substantial decline in our recognition performance with the application of the MLP classifier to our optimal geometric feature spaces. Conversely, the performance increase we achieved with the application of the MLP classifier to our spectral features was in agreement with the findings of [41]. The results, as a whole, showed a clear increase in GRF recognition performance when the MLP classifier was applied to feature spaces derived via the *unsupervised* PCA technique, and a clear decrease in performance when it was applied to feature spaces derived using *supervised* learning approaches. A likely explanation for these discrepancies points to the nature of the feature extraction techniques derivation. Each feature space was optimized to some degree using the KNN classifier, leading to an inherent performance *bias* toward the KNN algorithm used for feature space optimization. This bias would have been far greater in feature spaces derived via supervised learning, thus

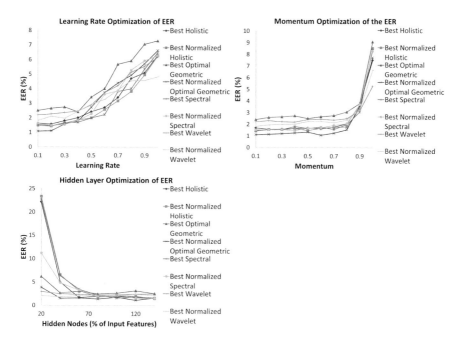

Fig. 6.6 MLP parameter optimization. This figure compares our best cross-validated EER values achieved with the optimization parameters used to achieve them, for all of our best performing preprocessing techniques. For any given parameter value, optimization was carried out by testing the given parameter value with every possible value for the other two parameters and returning the EER for the best combination

Table 6.2 MLP classifier performance

Optimal MLP classifier results

Preprocessor	ϵ	α	Hidden nodes	Thres.	CV EER (%)	EER incr. (%)
Best optimal geometric	0.4	0.1	29	0.05	2.41111	−80.8
Best normalized optimal Geometric	0.1	0.6	66	0.1656	1.07777	−506.2
Best holistic	0.2	0.2	21	0.1125	1.56666	38.6
Best normalized holistic	0.2	0.1	19	0.3	1.42222	30.4
Best spectral	0.2	0.1	19	0.0906	1.45555	28
Best normalized spectral	0.2	0.4	24	0.1062	1.47777	19.8
Best wavelet	0.1	0.4	64	0.2968	2.2	−70.6
Best normalized wavelet	0.1	0.1	39	0.2	1.82222	−65.6

This table demonstrates the best performance achieved by the MLP classifier for each preprocessing technique. The *threshold* shown is the threshold at which the EER value was calculated (a value between 0 and 1 derived from the *raw MLP output*) and the EER *improvement* represents improvement in recognition performance achieved by the optimal MLP variant over the results calculated in the previous two chapters outweighing any potential performance increases that would otherwise be achieved using the MLP classifier. Nevertheless, our MLP classifier overcame this inherent bias in the unsupervised feature spaces, which would seem to suggest that the MLP would lead to better recognition performance than the KNN classifier if it were to be used in place of the KNN classifier for feature space optimization.

6.3 Support Vector Machine

In the previous section we set out to demonstrate the claim that an MLP classifier could be used to derive *class-separating boundaries* not otherwise obtainable using a KNN classifier. The results obtained after applying the MLP classifier to our GRF data appeared to support this claim. However, research into MLPs has shown that they are prone to producing boundaries so tightly aligned with the specific characteristics of their training samples' class distribution that they may be unable to produce the strong *generalization* of boundaries needed to account for variability in the unseen test samples [20, 37]. One classification model that, in literature, has often been found to produce better class boundary generalization is the Support Vector Machine (SVM) [4]. The SVM-based classification technique was the second most widely used classification technique in our previously examined GRF recognition studies [27, 33, 34, 41], and in each of these studies it produced the best recognition performance when compared with all other studied classifiers. This *discriminative* classifier is an *eager* learner, in that it performs classification optimizations as part of a distinctive training phase; however, it could also be considered to have properties of an *instance-based* learning algorithm, in that it retains a subset of samples from the original training dataset to assist with the definition of boundaries during its usage. At the core of the SVM-based classification technique are one or more *binary linear classifiers* (SVMs), which, when grouped together, can be used to solve multi-class problems [25]. Each individual SVM is derived via solving for the *maximum-margin hyperplane* separating a provided two-class training dataset.

To understand the SVM *maximum-margin problem* consider the linear separator dividing the samples from two different classes in the two dimensional feature space, shown in Fig. 6.7. Although this linear separator could have been drawn in numerous locations and still have accomplished a separation of classes, the chosen location, which represents the *maximum distance* between the separator and nearest samples of either class, was selected because research has shown it often provides *strong generalization* boundaries [1]. Consequently, it follows that by solving for the line that forms this maximum-margin separator we would have a high likelihood of acquiring a set of boundaries that separate the samples of our different GRF subjects better than was possible with the classification methods of the previous two sections.

To solve for the maximum-margin separator, we first need to frame the aforementioned problem such that our linear separator is defined with respect to our *training samples*, which we will refer to as the feature vectors x_i for any sample i, and with respect to the *labels* of the two classes that we are attempting to divide, which we will identify as +1 for one class and −1 for the other ($y_i \in \{+1, -1\}$).

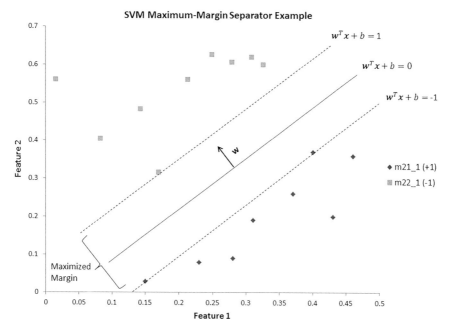

Fig. 6.7 SVM maximum-margin separator example. This figure demonstrates the separation of the samples for two different GRF subjects using the maximum-margin hyperplane $\boldsymbol{w}^T\boldsymbol{x}+\boldsymbol{b}=0$. In this example, the samples are represented in a 2-dimensional feature space, yet the separator is defined such that it would still be applicable for a space containing any number of dimensions

Looking again at Fig. 6.7 we can see that we have *bounded* the linear separator ($\boldsymbol{w}^T\boldsymbol{x}+b=0$.) between the two parallel lines representing the *margin boundaries* ($\boldsymbol{w}^T\boldsymbol{x}+b=1$ and $\boldsymbol{w}^T\boldsymbol{x}+b=-1$) that run through the points sitting nearest to the separator for both classes and satisfy the *constraints* in (6.9).

$$\boldsymbol{w}^T\boldsymbol{x}_i+b\geq 1 \quad \text{for} \quad y_i=+1$$
$$\boldsymbol{w}^T\boldsymbol{x}_i+b\leq -1 \quad \text{for} \quad y_i=-1 \tag{6.9}$$
$$\Rightarrow y_i\left(\boldsymbol{w}^T\boldsymbol{x}_i+b\right)\geq 1 \quad or \quad -y_i\left(\boldsymbol{w}^T\boldsymbol{x}_i+b\right)+1\leq 0 \quad \forall_i$$

At this point, we know the values for $\boldsymbol{x}$ as well as the values of $\boldsymbol{w}^T\boldsymbol{x}+b$ for the vector points that sit on the margin boundaries (referred to as the *support vectors*). If we were to calculate the distance from the separator to any point on either of the parallel margin boundaries we would find that the margin separating the two classes can be maximized by minimizing the *Euclidean norm* of our normal vector ($\|\boldsymbol{w}\|$), as demonstrated in (6.10). This *minimization problem*, bounded by the constraints defined in (6.9), is typically referred to as the primal optimization problem for SVM. By calculating the distance between our two margin lines h_1 and h_2 under the constraints of (6.9), it can be seen that when the Euclidean norm of the normal vector is minimized the margin will be maximized (Note: minimizing *half* of its square can improve computational efficiency without changing the result).

$$h_1: \boldsymbol{w}^T \boldsymbol{x_i} + b = 1, \quad h_2: \boldsymbol{w}^T \boldsymbol{x_i} + b = -1$$

$$\text{dist}\left(\boldsymbol{w}^T \boldsymbol{x_i} + b = 0, \boldsymbol{x_0}\right) = \frac{|\boldsymbol{w}^T \boldsymbol{x_0} + b|}{\|\boldsymbol{w}\|}, \quad \boldsymbol{x_0} \in h_1 \quad \vee \quad \boldsymbol{x_0} \in h_2$$

$$\Rightarrow \text{dist}\left(\boldsymbol{w}^T \boldsymbol{x_i} + b = 0, \boldsymbol{x_0}\right) = \frac{|1|}{\|\boldsymbol{w}\|}$$

$$\Rightarrow \text{dist}\left(\boldsymbol{w}^T \boldsymbol{x_i} + b = 1, \boldsymbol{w}^T \boldsymbol{x_i} + b = -1\right) = \frac{2}{\|\boldsymbol{w}\|} \text{ (SVM margin)}$$

$$(6.10)$$

Primal Optimization Problem (margin maximization) :

$$\min_{w,b} \frac{1}{2} \|\boldsymbol{w}\|^2 \quad \text{subject to} \quad -y_i\left(\boldsymbol{w}^T \boldsymbol{x_i} + b\right) + 1 \le 0 \qquad \forall_i$$

The Eqs. (6.9) and (6.10) demonstrated how the SVM maximum-margin problem can be presented as the *optimization* of a function bounded by constraints; alternatively, the problem can be redefined using *Lagrangian multipliers* ($\boldsymbol{\alpha}$) for the optimization of a single *unbounded* auxiliary function. Consider the Lagrangian *auxiliary function* defined in (6.11), which shows how Lagrangian multipliers (α) can be combined with the function being minimized and the left side of the constraint inequality from the optimization problem defined in (6.10); with l representing the number of different samples being optimized. If this function were to be maximized with respect to its Lagrangian multipliers then the resulting optimization would be found to produce an *infinitely* large value when the original constraints are not satisfied, but would produce a value equivalent to *objective function* ($\frac{1}{2} \|\boldsymbol{w}\|^2$) when the original constraints are satisfied (see (6.12)). In other words, the part of the auxiliary function derived from the Primal Optimization Problem constraints may become infinitely large when the constraints are not satisfied (i.e., $-y_i(\boldsymbol{w}^T \boldsymbol{x_i} + b) + 1 > 0$); however, when they are satisfied the maximum value for the constraint part of the auxiliary function will be 0, thus giving $\mathcal{L}_P(\boldsymbol{w}, b)$, the value of the function being optimized in (6.10). Knowing this, it can be shown that by *minimizing* the function in (6.12) with respect to w and b (finding $\min_{w,b} \mathcal{L}_P(\boldsymbol{w}, b)$) we end up with a problem with an *equivalent solution* to that of the primal optimization problem. This form of the optimization problem is often referred to as the *Lagrangian primal problem*, and, with a bit of extra derivation it can also be shown that there is a *dual* representation of this Lagrangian problem which further simplifies the computation required to find the solution to the primal optimization problem [7].

$$\mathcal{L}(\boldsymbol{w}, b, \boldsymbol{\alpha}) = \frac{1}{2} \|\boldsymbol{w}\|^2 - \sum_{i=1}^{l} \alpha_i y_i \left(\boldsymbol{w}^T \boldsymbol{x_i} + b\right) + \sum_{i=1}^{l} \alpha_i \qquad (6.11)$$

$$\mathcal{L}_P(\boldsymbol{w}, b) = \max_{\boldsymbol{\alpha}: \alpha_i \ge 0} \mathcal{L}(\boldsymbol{w}, b, \boldsymbol{\alpha})$$

$$= \begin{cases} \frac{1}{2} \|\boldsymbol{w}\|^2 & \text{if } \boldsymbol{w} \text{ satisfies the primal constraints} \\ \infty & \text{otherwise} \end{cases} \qquad (6.12)$$

The *dual form* of the Lagrangian problem (or *Lagrangian dual problem*) derives from the concept of *duality*, whereby there exist two related optimization problems, a *primal* and a *dual* problem, with the solution to the dual problem forming the *lower bound* to the solution of the primal problem. The solution to the dual problem always forms the lower bound to the primal problem; however, in certain circumstances this lower bound may in fact be a common *shared* optimal solution to both problems. In the SVM case, when the relationship between the primal (p^*) and dual (d^*) problems in (6.13) is considered, it can be shown that these problems satisfy the necessary *conditions* for equality in their solution [7]. Thus solving the dual problem will also reveal the solution to the maximum-margin separator of the primal problem. Moreover, because the auxiliary function that is being optimized (6.11) is *convex* with respect to w and b for any fixed value of α, by taking the *partial derivatives* of w and b at the optimal value, it is possible to derive the properties shown in (6.14). These relationships are derived by minimizing $\mathcal{L}(w, b, \alpha)$ with respect to w and b, and the *optimal minimum* value occurs when the partial derivatives are equal to *zero* because the function $\mathcal{L}(w, b, \alpha)$ is convex with respect to both parameters for any value of α.

$$d^* = \max_{\alpha : \alpha_i \geq 0} \min_{w,b} \mathcal{L}(w, b, \alpha) \quad \leq \quad \min_{w,b} \max_{\alpha : \alpha_i \geq 0} \mathcal{L}(w, b, \alpha) = p^* \tag{6.13}$$

$$\frac{\partial}{\partial w} \mathcal{L}(w, b, \alpha) = w - \sum_{i=0}^{l} \alpha_i y_i x_i = 0 \Rightarrow w = \sum_{i=0}^{l} \alpha_i y_i x_i$$

$$\frac{\partial}{\partial b} \mathcal{L}(w, b, \alpha) = \sum_{i=1}^{l} \alpha_i y_i = 0 \tag{6.14}$$

Using the properties discovered in (6.14), the SVM Lagrangian dual problem's auxiliary function can be simplified such that it only needs to be optimized for the *single* variable α, as shown in (6.15). In this *simplified* SVM Lagrangian dual problem the findings from (6.14) have been substituted into the auxiliary function of (6.11), revealing a form of the maximum-margin problem that can easily be solved via *quadratic optimization* [3].

$$\mathcal{L}_D(\alpha) = \min_{w,b} \mathcal{L}(w, b, \alpha) = \frac{1}{2} \left(\sum_{i=1}^{l} \alpha_i y_i x_i \right)^T \left(\sum_{j=1}^{l} \alpha_j y_j x_j \right)$$

$$- \sum_{i=1}^{l} \alpha_i y_i \left(\sum_{j=1}^{l} \alpha_j y_j x_j \right)^T x_i - b \sum_{i=1}^{l} \alpha_i y_i + \sum_{i=1}^{l} \alpha_i$$

$$= \sum_{i=1}^{l} \alpha_i - \frac{1}{2} \sum_{i,j=1}^{l} \alpha_i \alpha_j y_i y_j x_i^T x_j \tag{6.15}$$

$$d^* = \max_{\alpha} \mathcal{L}_D(\alpha) \quad \text{subject to} \quad \alpha_i \geq 0,$$

$$i = 1, \dots, l \quad \text{and} \quad \sum_{i=1}^{l} \alpha_i y_i = 0$$

From this point it is then possible to derive the value of b for the maximum-margin separator as the *midway point* between the values it takes on the two parallel margins, as shown in (6.16). This can be done using the discovered values of w. In this equation the *maximum* value for points identified by $y_i = -1$ will sit on one margin while the *minimum* value for those identified by $y_i = 1$ will sit on the other, so by taking one-half of the combined b component for each margin it is possible to obtain its value at the separator.

$$b = -\frac{\max\limits_{i:y_i=-1} w^T x_i + \min\limits_{i:y_i=1} w^T x_i}{2} \tag{6.16}$$

Finally, the derived parameters can be substituted into (6.17) to form the desired SVM *binary classifier* (the foundation of the multi-class SVM classification technique). In this equation the maximum-margin separator is shown as it would be used to perform classification for a given vector point x and training points $1 \ldots l$. If the resulting value was greater than *zero* then the vector x would be classified as belonging to the class which was given the label $y_i = 1$, otherwise it is classified as belonging to the class given the label $y_i = -1$. Moreover, because it is also possible to show that the SVM dual form problem satisfies the *Karush–Kuhn–Tucker* complementary condition [4], it can also be shown that $\alpha_i = 0$ for any training vector point x_i that does not fall on the margin; thus, when performing classification via this equation, classifier *memory usage* can often be optimized, as only support vectors (typically a small subset of training samples) are needed to derive the classification boundary and all other values are *ignored*.

$$w^T x + b = \left(\sum_{i=1}^{l} \alpha_i y_i x_i \right)^T x + b = \sum_{i=1}^{l} \alpha_i y_i x_i^T x + b \tag{6.17}$$

The SVM classifier as it has been defined to this point has a couple of *weaknesses* that make it poorly suited to more complicated classification tasks, namely it depends upon its training data being *linearly* separable and it can be susceptible to undesirable influence from *outlier points*. To address these problems a slightly different variant of SVM known as *soft margin* SVM can be used. In soft margin SVM training samples are allowed to sit *within* the margin or even on the *wrong side* of the separator. This is accomplished by redefining the margin constraint as $1 - \xi_i$, where the nonnegative *slack* variables ξ_i represent the *degree* to which a given vector point x_i is sitting on the wrong side of its class' margin boundary. In this case, the primal optimization problem becomes the simultaneous *maximization* of the margin separating the two classes and *minimization* of the degree to which training points are situated on the incorrect side of the margin to achieve the respective maximum margin. The formal definition of the soft margin SVM's primal optimization problem is demonstrated in (6.18). In this equation samples are allowed to sit on the wrong side of their margin boundaries; however, in doing so the degree to which each is misclassified ($\xi_i \geq 0$) incurs a *penalty cost* in the minimization of the objective function. The *regularization* parameter C is used to adjust the *weight* of which the penalty is applied; in other words, allowing for adjustments to the relative weighting given to the two problem components.

$$\min_{w,\xi,b} \frac{1}{2}\|w\|^2 + C\sum_{i=1}^{l}\xi_i \tag{6.18}$$
$$\text{subject to } -y_i(wx_i + b) + (1 - \xi_i) \leq 0 \quad \forall_i$$

From here, using an additional Lagrangian multiplier ($\beta \geq 0$), the Lagrangian function of (6.11) can be redefined to reflect the inclusion of the new slack variables as shown in (6.19). This equation demonstrates the Lagrangian representation of the soft margin SVM objective function. It is similar to (6.11); however, it includes the addition of the slack variables (ξ) and their respective Lagrangian multipliers (β).

$$\mathcal{L}(w, b, \xi, \alpha, \beta) = \frac{1}{2}\|w\|^2$$
$$+ C\sum_{i=1}^{l}\xi_i - \sum_{i=1}^{l}\alpha_i\left(y_i\left(w^T x_i + b\right) - (1 - \xi_i)\right) - \sum_{i=1}^{l}\beta_i\xi_i \tag{6.19}$$

To get the *dual form* of this new problem the *partial derivative* of the new Lagrangian function can be taken with respect to ξ, in addition to the partial derivatives of w and b shown earlier in (6.14); this in turn results in a new *dual problem* (6.20) that only differs from the Lagrangian dual problem previously defined in (6.15) in the definition of the *constraint* placed on α. Under this new problem definition, the support vectors include not only the trial sample points that sit *on* the margins, but also trial samples that sit on the *wrong side* of the margins.

$$\frac{\partial}{\partial \xi}\mathcal{L}(w, b, \xi, \alpha, \beta) = -\sum_{i=1}^{l}\alpha_i + C - \sum_{i=1}^{l}\beta_i$$
$$\Rightarrow \alpha_i = C - \beta_i \quad \forall_i \quad \alpha_i \geq 0, \beta_i \geq 0$$
$$\Rightarrow C - \alpha_i \geq 0 \text{ and } \alpha_i \geq 0 \Rightarrow \alpha_i \leq C \text{ and } \alpha_i \geq 0$$
$$\mathcal{L}_D(\alpha) = \min_{w,\xi,b}\mathcal{L}(w, b, \xi, \alpha, \beta) = \sum_{i=1}^{l}\alpha_i - \frac{1}{2}\sum_{i,j=1}^{l}\alpha_i\alpha_j y_i y_j x_i^T x_j \tag{6.20}$$
$$d^* = \max_{\alpha}\mathcal{L}_D(\alpha) \quad \text{subject to} \quad 0 \leq \alpha_i \leq C,$$
$$i = 1,\ldots,l \quad \text{and} \quad \sum_{i=1}^{l}\alpha_i y_i = 0$$

Additionally, when using the *soft margin* form of SVM, the new technique shown in (6.21) must be used to solve for b, as opposed to the technique previously demonstrated in (6.16). In this new equation the solution to b for soft margin SVM is shown with S being the subset of training points that form the *support vectors* (i.e., the points sitting on the margin boundaries together with those sitting on the wrong side of the margin boundaries). Yet, having solved for its parameters, SVM classification can once again be accomplished using (6.17) via substituting in the new soft margin solutions for α and b.

$$b = \frac{1}{|S|} \sum_{i \in S} \left(y_i - \sum_{j \in S} \alpha_i y_i x_i^T x_j \right) \tag{6.21}$$

The application of soft margin SVM can be useful when solving problems that are nearly, but not quite, linearly separable, yet it will fail when applied to problems that are clearly *not* linearly separable, such as the one on the left side of Fig. 6.8. To get around this limitation SVMs use a technique known as the *kernel trick*, which is premised on transforming a nonlinearly separable feature space into a *higher dimensional* feature space that can more easily be separated linearly (for instance the feature space on the *right* side of Fig. 6.8). Under the kernel trick the function $\emptyset(x)$ is defined as a function that transforms a sample x into a higher dimensional feature space. If this function were applied to the two sample points in the Lagrangian dual problem of (6.20) then, rather than searching for a linear separator in our original feature space, we would instead be searching for a linear separator in the high-dimensional space defined by $\emptyset$. In doing so the application of the dimension expansion function would result in the *dot product* of the original dual problem equation $x_i^T x_j$ becoming $\emptyset(x_i)^T \emptyset(x_j)$, which in many cases can be efficiently computed *without* ever having to compute $\emptyset$ for x_i and x_j individually [4]. The function

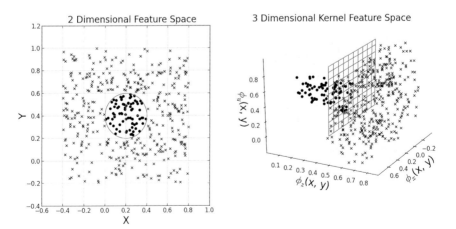

Fig. 6.8 Kernel space transformation example. The two diagrams in this figure demonstrate how a dataset that cannot be linearly separated in 2-dimensions (the diagram on the *left*) can be transformed into a higher dimensional kernel space that can be separated linearly (shown in the diagram on the *right*)

that accomplishes this simplified computation of the dot product over a higher dimensional space is known as the *kernel function* and is typically denoted as $\kappa(x_i, x_j)$. To use the kernel trick with soft margin SVM, the chosen kernel function would be substituted to *replace* of any dot products of the *input* samples x_i and x_j within the Lagrangian dual problem (6.20), the calculation of b (6.21), and the binary SVM classifier (6.17). As for the definition of the kernel function itself, there are many valid forms its equation could take, but research has shown that one kernel in particular, the *Gaussian Radial Basis Function* (RBF) kernel (6.22), outperforms most others when applied to the GRF-based gait recognition problem [34]. This equation takes two point vectors as input and produces output equivalent to that which would be produced if the dot product was taken after both points were transformed into a higher dimensional space. The shape of the transform can be tuned to *optimize* the strength of the SVM separator by adjusting the kernel parameter γ.

$$\kappa(x_i, x_j) = \emptyset(x_i)^T \emptyset(x_j) = \exp\left(-\gamma \|x_i - x_j\|^2\right) \qquad (6.22)$$

Using a single soft margin kernel SVM gives us a powerful tool for classifying the points in *two-class* problems; however, to perform SVM classification on any *multi-class* problem, including our GRF subject recognition problem, we require multiple SVMs and a strategy to train them and evaluate their results. Two strategies are commonly used to accomplish multi-class classification with SVM: *one-against-one* and *one-against-all*. In the typical *one-against-one* strategy individual SVMs are first trained for each of the $C(C-1)/2$ training class pairs taken from the C training classes, then, during classification the class that each SVM selects is given single *vote* and the sample being classified is classified as belonging to the class that gains the most votes; this strategy was used for GRF recognition in [27]. A slightly different approach is taken in the typical *one-against-all* strategy; rather than deriving SVMs for every class pair, in one-against-all an SVM is first trained for each class against a grouping of *all* the samples in *every* other training class, then during classification the class with the highest raw SVM *output value* is assigned to the sample being classified. In their common form neither of the two multi-class SVM strategies output *posterior probabilities*, however, both can be manipulated to give such output [25]. A common approach, implemented in the popular *LIBSVM* tool [6], uses a modified version of the one-against-one strategy to calculate posterior probabilities. In this case, rather than treating the SVM's as binary classifiers with votes, the *unscaled raw output* of each individual SVM output is passed into a *sigmoid function* to produce an estimate of the *pairwise probabilities* for each of the $C(C-1)/2$ training class pairs. The sigmoid function demonstrated in (6.23) estimates the probability that a given sample x belongs to class i for the output of an SVM trained across classes i and j. To estimate the values for A and B in this equation the chosen implementation minimizes the *negative log likelihood* of the training data [22]. Having found the *pairwise probabilities*, the probability (p_i) that a given tested sample x belongs to a given training class i can be estimated by solving for the optimization problem in (6.24), as discussed by Wu et al. in [45]. This equation shows how the pairwise probabilities (r) calculated in (6.23) can be incorporated into an optimization problem to solve for the probability (p) that the sample used to generate the pairwise probabilities belongs to each of the training data classes.

$$r_{ij} \approx P(y = i | y = i \, or \, j, \boldsymbol{x}) \approx \frac{1}{1 + \exp\left(Af_{ij}(\boldsymbol{x}) + B\right)} \qquad (6.23)$$

$$\min_{\boldsymbol{p}} \frac{1}{2} \sum_{i=1}^{k} \sum_{j:j=1, j \neq i}^{k} \left(r_{ji} p_i - r_{ij} p_j\right)^2 \qquad (6.24)$$

$$\text{subject to} \quad p_i \geq 0, \, \forall_i, \quad \sum_{i=1}^{k} p_i = 1$$

To this point we have demonstrated the *probabilistic multi-class soft margin* variant of *kernel SVM* previously used in [41]. This is the configuration we decided upon using for our own implementation of SVM for GRF recognition, which we developed using the *Encog* [14] wrapper of the popular *LIBSVM* tool [6]. For our kernel function, we opted for the *Gaussian RBF* function as this was selected as the kernel function in all of the previously SVM-based GRF recognition studies. To effectively generate EER values using this technique we created an *extension* of the standard Encog SVM class, constructing it to initialize with the LIBSVM *probability* parameter flag set to 1. Furthermore, we altered the default LIBSVM probability generation behavior, which, due to the *pseudorandom* cross-validation used in its probability generation algorithm, effectively returned *nondeterministic* probabilities. In our implementation we had the LIBSVM code use deterministic probabilities by setting Encog's *SupportClass.Random* value to use a seeded random number rather than the default pseudorandom number. Finally, we optimized our SVM classifier with respect to its input parameters: the features passed in for training and testing, the value of the *regularization* parameter (C), and the value of the *kernel* parameter (γ). First, to reduce the undesirable bias from any particular input feature we *rescaled* all our training and testing input features to fall within the range of [0, 1], as defined by the respective minimum and maximum values for each feature within the training dataset. Having rescaled our input features, the values of our two additional SVM parameters were then optimized via an *exhaustive search* of various parameter combinations, similar to the approach used in our MLP optimization. By picking a reasonable distribution of arbitrary parameter values we were able to identify the *range* of values for each parameter that optimized the GRF recognition EER results when used with each of our preprocessor configurations (shown in Fig. 6.9).

In Fig. 6.9 we can clearly see that the regularization parameter is ineffective for GRF recognition when its value is below about 0.25, thereafter having little *influence* on performance when given larger values. Conversely, the optimization of the kernel parameter achieved its best performance when given a value below 1 and

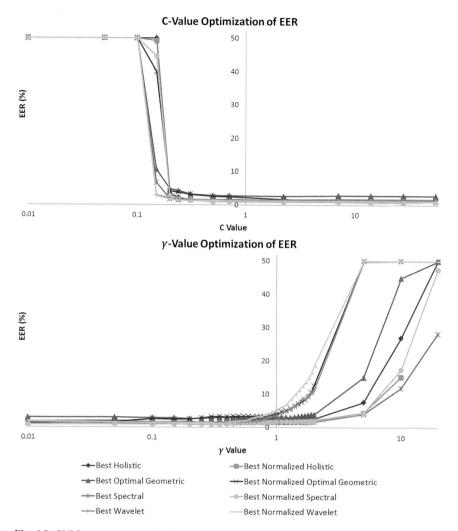

Fig. 6.9 SVM parameter optimization. This figure compares our best cross-validated EER values achieved with the SVM classifier to the optimization parameters used to achieve them for all of our best performing preprocessing techniques. For both given parameter values optimization was carried out by testing the given parameter value with every possible value for the other parameter and returning the EER for the best combination

quickly became ineffective when given larger values. The combination of parameters that produced the best GRF recognition results across our best feature extraction configurations are shown in Table 6.3. As was the case with our MLP classifier, the use of an SVM classifier led to a substantial decrease in GRF recognition performance when used in combination with our optimal geometric and wavelet feature spaces, likely owing to their relatively strong inherent *performance bias* toward KNN (a result of the use of KNN in our feature extractor optimization).

Table 6.3 SVM classifier performance

Optimal SVM classifier results					
Preprocessor	C	γ	Thres.	CV EER (%)	EER incr. (%)
Best optimal geometric	2.24	0.2	0.1531	2.61111	−95.8
Best normalized optimal geometric	56.56	0.01	0.2109	1.04444	−487.5
Best holistic	2.24	0.1	0.1656	0.96666	62.1
Best normalized holistic	0.71	0.25	0.1531	1.38888	32
Best spectral	7.07	0.05	0.1859	1.31111	35.1
Best normalized spectral	14.14	0.01	0.1984	0.83333	54.8
Best wavelet	2.24	0.15	0.1593	1.43333	−11.2
Best normalized wavelet	0.71	0.35	0.1406	1.2	−9

This table demonstrates the best performance achieved by the SVM classifier for each preprocessing technique. The *threshold* shown is the threshold at which the EER value was calculated (a value between 0 and 1 derived from the *raw posterior probability output*) and the EER *improvement* represents the improvement in recognition performance achieved by the optimal SVM variant over the results calculated in the previous two chapters

However, as was also the case with our MLP classifier, the GRF recognition performance increased for our two feature spaces that were less subject to KNN bias (our holistic and spectral feature spaces). In fact the use of the SVM classifier together with our holistic and spectral feature spaces actually led to a significant increase in GRF recognition performance when compared with the MLP classification results, most notably in the *non-normalized* holistic and *normalized* spectral feature spaces. These findings, when accounting for the KNN bias, support the findings of previous GRF recognition studies [27, 33, 34, 41], which found SVM improved recognition performance over the KNN classifier. Furthermore, our findings on the comparison of SVM with MLP reflected those of [41], with SVM generally achieving comparable or better GRF recognition performance than MLP across a variety of different feature spaces.

6.4   Linear Discriminant Analysis

In Sect. 3.4 of Chap. 3 we discussed the categorization of classifiers as following either a *generative* or *discriminative* model for establishing whether a given sample belongs to a specific class. The classifiers we have examined so far were all categorized as being discriminative classifiers because their posterior class probabilities were derived *directly* from their optimized outputs, without regard for the underlying *class conditional densities*. In this section we explore the use of the *eager* learning classifier known as *Linear Discriminant Analysis* (LDA), which, despite its name, takes a *generative* approach to modeling posterior class probabilities. Traditionally, in its basic form, this classifier assumes each class has a *Gaussian* distribution of members with a common degree of variance across all classes [13]. Under such assumptions posterior probabilities are derived by first finding the position of a tested sample with respect to each class' conditional multivariate *Gaussian probability density function*, then using *Bayes' rule* to estimate the posterior class probabilities. The multivariate Gaussian probability density function, shown in (6.25), is used for estimating the *likelihood* of the vector sample x in an m-dimensional feature space, given the class labelled y_i and a vector *class mean* μ_{y_i} (estimated using training data). Under LDA, a common *covariance matrix* Σ is estimated using training data across all classes; this differs from the alternative *Quadratic Discriminant Analysis* (QDA) [39], which replaces Σ with class-dependent covariance matrices (Σ_i). Using *Bayes' rule* (6.26) the probability estimated in (6.25) can be used to derive the probability that the class y_i corresponds to the given sample x; however, doing so also requires an a priori estimate of the likelihood that a sample of class y_i would appear as an input ($P(y_i)$) for all k classes.

$$P(x|y_i) \approx \frac{1}{(2\pi)^{m/2}|\Sigma|^{1/2}} \exp\left(\frac{1}{2}\left(x - \mu_{y_i}\right)\Sigma^{-1}\left(x - \mu_{y_i}\right)\right) \qquad (6.25)$$

$$P(y_i|x) = \frac{P(x|y_i)P(y_i)}{P(x)} = \frac{P(x|y_i)P(y_i)}{\sum_{j=1}^{k} P(x|y_j)P(y_j)} \qquad (6.26)$$

The aforementioned LDA technique also comes in a *reduced* form which involves first using *Fisher's dimensionality reduction* technique to project the training and testing samples down to some smaller set of dimensions maximizing the statistical separation of classes, and then using this more discriminative reduced feature space to perform the traditional LDA classification [13]. Hastie et al. have suggested that the reduced form of LDA can produce better classification performance than the nonreduced form [13]; however, it is also susceptible to a deficiency known as the *small sample size* (SSS) problem [17], which occurs when the number of dimensions in the feature set being classified is greater than the number of data samples used for training. With regards to previous GRF recognition studies, only [27] previously used LDA for GRF recognition and no details were given on the actual variant of LDA used. For the purpose of our research we have elected to study the GRF recognition performance of a non-SSS susceptible variant of reduced LDA known as *Uncorrelated* LDA (ULDA) [46] together with the *kernelized* variation on this technique known as *Kernelized Uncorrelated Discriminant Analysis* (KUDA) [42].

When examining the LDA *class conditional probability* definition in (6.25) it becomes apparent that the single discriminating factor in LDA is the *squared Mahalanobis distance*, which forms its *exponent* term. When posterior probabilities are not required and the a priori likelihood of each class is considered to be *equal*, classification can be accomplished simply by using a *maximum likelihood estimator* of this distance [46]. Yet, a consequence of this dependency on a single feature space spanning distance metric is the erosion of the LDA classifier's discriminative ability when classification is performed on feature spaces with a significant number of weakly discriminant features; in this case the weak features would mask the stronger features, resulting in decreased classification performance. The *reduced* form of LDA mitigates this problem by reducing the *dimensionality* of the feature space using the *supervised* dimensionality reduction technique proposed by Fisher [2]. This technique, which is often also referred to as LDA or *Fisher Discriminant Analysis* (FDA) when used for feature extraction [16], is similar to PCA (Sect. 4.2 of Chap. 4) in that it uses *Eigen Decomposition* to derive a dimensionality reducing transformation matrix, but differs in metric around which the dimensionality reduction is optimized. Under PCA, the *zeroed* covariance matrix is taken as the *optimization criterion* and the dimensionality reducing transformation matrix is acquired by performing Eigen Decomposition on this criterion and extracting the subset of the eigenvectors corresponding to the largest eigenvalues returned by the decomposition; the result of this is a dimensionality reduction that *preserves* as much *variance* as possible. In contrast, the dimensionality reduction for reduced LDA is based around an optimization criterion known as the *Fisher criterion* (6.27). Solving for this criterion allows for a reduction in dimensionality that *maximizes* the separation of *class means* (the *between*-class scatter matrix S_b), while simultaneously *minimizing* the *degree* of *variance* across all classes (the *within*-class scatter matrix S_w), thus producing a dimensionality reduction that preserves as much of a separation between each class's *Gaussian norms* as possible. In (6.28) we see the definition of these *pooled within* (S_w) and *between* (S_b) class scatter matrices that are essential to performing reduced LDA. As demonstrated in the equation, they can be represented as the *square* of two other matrices (their *half-forms*), with A being the set of samples, A_i being the set of samples associated with class i, c containing the mean vector for each feature across all k classes, $c^{(i)}$ containing the mean vector across class i, n being the total number of samples, and n_i being the total number of samples in class i.

$$W = \underset{W}{\operatorname{argmax}} \, tr\left(\frac{W^T S_b W}{W^T S_w W}\right) \tag{6.27}$$

$$c = \frac{1}{n}Ae, \quad e = (1, 1, \ldots, 1)^T \in R^n$$

$$c^{(i)} = \frac{1}{n_i}A_i e^{(i)}, \quad e^{(i)} = (1, 1, \ldots 1)^T \in R^{n_i}$$

$$H_w = \frac{1}{\sqrt{n}}\left[A_1 - c^{(1)}\left(e^{(1)}\right)^T, \ldots, A_k - c^{(k)}\left(e^{(k)}\right)^T\right] \tag{6.28}$$

$$H_b = \frac{1}{\sqrt{n}}\left[\sqrt{n_1}\left(c^{(1)} - c\right), \ldots, \sqrt{n_k}\left(c^{(k)} - c\right)\right]$$

$$S_w = H_w H_w^T, \quad S_b = H_b H_b^T$$

With a bit of work the Fisher criterion can be reformulated as a problem that can be solved via Eigen Decomposition, having the *transformation matrix* G defined as the *row-appended* eigenvectors corresponding to the *nonzero* eigenvalues (6.29). In this case W will be formed by the *eigenvectors* associated with the q nonzero

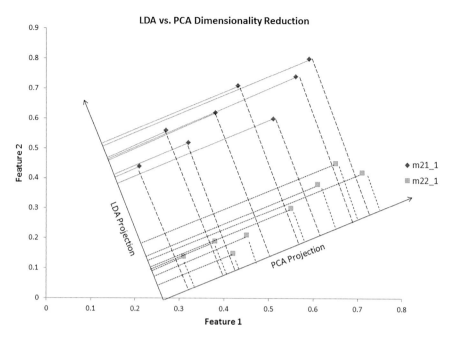

Fig. 6.10 LDA versus PCA dimensionality reduction. This figure compares the dimensionality reduction in PCA, which projects data into the dimension of highest variance, with the reduced LDA dimensionality reduction, which projects data into the dimension maximizing the distance between class distributions. In this case the projection into the LDA dimension produces a result that has the two classes separated, while the projection into the PCA dimension has the data from the two classes interspersed

eigenvalues in the *diagonal* eigenvalue matrix $\mathbf{\Lambda}$. Graphically, the discriminative properties provided by the Fisher technique over PCA become particularly apparent when the dataset variance runs *perpendicular* to the separation of classes, as demonstrated in Fig. 6.10.

$$\frac{d}{d\mathbf{W}}\frac{\mathbf{W}^T S_b \mathbf{W}}{\mathbf{W}^T S_w \mathbf{W}} = \frac{\frac{d}{d\mathbf{W}}\left(\mathbf{W}^T S_b \mathbf{W}\right)\mathbf{W}^T S_w \mathbf{W} - \frac{d}{d\mathbf{W}}\left(\mathbf{W}^T S_w \mathbf{W}\right)\mathbf{W}^T S_b \mathbf{W}}{\left(\mathbf{W}^T S_w \mathbf{W}\right)^2}$$

$$\Rightarrow \underset{\mathbf{W}}{\mathrm{argmax}}\, tr\left(\frac{\mathbf{W}^T S_b \mathbf{W}}{\mathbf{W}^T S_w \mathbf{W}}\right) = \frac{(2S_b \mathbf{W})\mathbf{W}^T S_w \mathbf{W} - (2S_w \mathbf{W})\mathbf{W}^T S_b \mathbf{W}}{\left(\mathbf{W}^T S_w \mathbf{W}\right)^2} = 0$$

$$\Rightarrow \mathbf{W}^T S_w \mathbf{W}(S_w \mathbf{W}) - \mathbf{W}^T S_b \mathbf{W}(S_w \mathbf{W}) = 0$$

$$\Rightarrow \frac{\mathbf{W}^T S_w \mathbf{W}(S_b \mathbf{W})}{\mathbf{W}^T S_w \mathbf{W}} - \frac{\mathbf{W}^T S_b \mathbf{W}(S_w \mathbf{W})}{\mathbf{W}^T S_w \mathbf{W}} = S_b \mathbf{W} - \frac{\mathbf{W}^T S_b \mathbf{W}}{\mathbf{W}^T S_w \mathbf{W}}(S_w \mathbf{W}) = 0$$

$$\Rightarrow S_b \mathbf{W} = \mathbf{\Lambda}(S_w \mathbf{W}) \quad or \quad S_w^{-1} S_b \mathbf{W} = \mathbf{\Lambda}\mathbf{W},$$

$$(6.29)$$

where $\mathbf{\Lambda} = \frac{\mathbf{W}^T S_b \mathbf{W}}{\mathbf{W}^T S_w \mathbf{W}}$ and $\mathbf{G} = \mathbf{W}_q$

Unlike PCA, which, prior to reduction, will produce as many dimensions as there are dimensions of variance in a dataset, the Fisher dimensionality reduction technique is *bounded* by the *number of classes* identified in its training dataset. This owes to the fact that a dataset with C unique classes will result in a matrix S_b having a *rank* with an upper bounding of $C - 1$, and therefore, the Eigen Decomposition of (6.29) can only ever produce at most $C - 1$ nonzero eigenvalues [46]. Moreover, in datasets containing *fewer* dimensions than classes this bound on the number of dimensions returned would instead be equal to the *number of dimensions* in the dataset, thus the upper bound on the number of dimensions produced for reduced LDA can be represented as the *minimum* of the two possible boundaries. This *upper bound* for dimensionality reduction via the Fisher dimensionality reduction technique over a dataset with p dimensions and C different classes is shown in (6.30); in this equation x represents a sample in the dataset while $G^T x$ represents the transformation of x into the reduced dimensional space. The *lower boundary*, in contrast, has no such restrictions and, in using this technique, any dataset could be reduced down to its single most discriminant dimension; however, for the purpose of our research the reduced LDA classifier is always assumed to use a dimensionality reduction equal to its upper bound (note that for datasets with a large number of classes this reduction may not be small enough). With a chosen dimensionality reduction strategy in place, once the dimensionality reducing *transformation matrix* (*G*) has been found, the *class conditional probability density function* for reduced LDA classifier will take the form shown in (6.31); in this case, each dataset-dependent input is multiplied with the transformation matrix *G*.

$$m = \max(p, C - 1) \quad \text{with} \quad x \epsilon R^p \quad \text{and} \quad G^T x \in R^m \tag{6.30}$$

$$\tilde{x} = G^T x, \quad \tilde{\mu}_{y_i} = G^T \mu_{y_i}, \quad \tilde{S}_w = G^T S_w G$$

$$P(x|y_i) \approx \frac{1}{(2\pi)^{m/2} |\tilde{S}_w|^{1/2}} \exp\left(\frac{1}{2} \left(\tilde{x} - \tilde{\mu}_{y_i} \right) \tilde{S}_w^{-1} \left(\tilde{x} - \tilde{\mu}_{y_i} \right) \right) \tag{6.31}$$

One unfortunate *weakness* in the (6.31) formulation for reduced LDA is the strict requirement that the *within*-class scatter matrix be *invertible*. The introduction of a non-invertible *singular* within-class scatter matrix is a common problem for datasets that contain more samples than dimensions (the SSS problem). To get around this limitation a number of techniques have been proposed including Regularized LDA [9], Nullspace LDA [24], and Penalized LDA [12] among others. Among the proposed solutions, the ULDA technique described by Ye in [46] provides a convenient *generalization* to the reduced LDA for situations with both *singular* and *non-singular* within-class scatter matrices. Under the ULDA technique, the *inverted term* in the optimization criterion is reformulated using the *pseudoinverse*, giving ULDA the valuable property of being equivalent to the standard reduced LDA technique when dealing with the case of a non-singular scatter matrix in the criterion's denominator (in this case the pseudoinverse is equal to the actual inverse). Moreover, when dealing with the case of a singular scatter matrix ULDA acts as an *extension* on reduced LDA, using the pseudoinverse's optimal inverse solution in the case for which an inverse could not be found using standard matrix inversion [44]. The ULDA technique also takes advantage of the fact that the Fisher optimization criterion can be solved using an equivalent form that has the *total scatter matrix* (6.32) in the place of the *within-class scatter matrix* [23]; in this new formulation of the total scatter matrix, we have A as the dataset, c being the vector of dataset means, and e being the vector of *1s* defined in (6.28). With these adaptations to the standard reduced LDA, the ULDA *optimization criterion* takes on the new form shown in (6.33). Note that, unlike the Fisher Optimization Criterion, this criterion has the total scatter matrix (S_t) in the place of the within-class scatter matrix, and uses the pseudoinverse in the place of the inverse.

$$H_t = \frac{1}{\sqrt{n}}\left(A - ce^T\right), \quad S_t = H_t H_t^T \tag{6.32}$$

$$G = \arg\max_G \operatorname{tr}\left(\left(G^T S_t G\right)^+ \left(G^T S_b G\right)\right) \tag{6.33}$$

The efficient ULDA algorithm that Ye [46] proposed is premised on solving for the matrix X that simultaneously *diagonalizes* the three different scatter matrices (S_b, S_w, and S_t). Ye was able to show that this could be done using only the *half-between*-class scatter matrix (H_b) and *half- total* class scatter matrix (H_t); in Ye's algorithm these *half scatter matrices* are assumed to have *rows* corresponding to *dimensions* and *columns* corresponding to individual data *samples*, the *transpose* of the form normally taken by these matrices. The first step in the algorithm involves using *Singular Value Decomposition* (SVD) to *decompose* the half-total scatter matrix into its *orthogonal* and *diagonal* components. The *full* SVD of H_t, shown in (6.34), decomposes it into two *orthogonal* components U and V as well as a *diagonal* component Σ. Here the previously defined half-scatter matrices have been transposed so that each row represents one of the m dimensions in the feature space and each column represents one of the n samples in the training dataset.

$$H_t = U\Sigma V^T,$$

with $\quad \Sigma = \begin{pmatrix} \Sigma_t & 0 \\ 0 & 0 \end{pmatrix}, \quad \Sigma_t \in R^{t\times t}, \quad$ and $\quad t = \operatorname{rank}(S_t), \tag{6.34}$

$$H_t \in R^{m\times n}, \quad U \in R^{m\times m}, \quad \Sigma \in R^{m\times n}, \quad V \in R^{n\times n}$$

When accounting for the relationship between the scatter matrices it was shown that a large part of the SVD can be *ignored* and this first step could simply be accomplished using the more efficient *reduced* (or economy) SVD technique [28]. The derivation in (6.35) demonstrates that *reduced SVD*, which involves calculating only the first t columns of U, can be used in the place of *full SVD* in modeling the relationship between the scatter matrices. In this equation, t continues to represent the *rank* of the total scatter matrix and the fact that all matrices involved are *positive semidefinite* means only the sum of $U_1^T S_b U_1$ and $U_1^T S_w U_1$ will be *nonzero*, with a sum of Σ_t^2.

$$S_t = U\Sigma V^T V \Sigma^T U^T = U\Sigma\Sigma^T U^T = \begin{pmatrix} \Sigma_t^2 & 0 \\ 0 & 0 \end{pmatrix},$$

$$U = (U_1, U_2), \quad U_1 \in R^{m \times t} \text{ and } U_2 \in R^{m \times (m-t)}$$

$$\begin{pmatrix} \Sigma_t^2 & 0 \\ 0 & 0 \end{pmatrix} = U^T S_t U = U^T (S_b + S_w) U$$

$$= \begin{pmatrix} U_1^T \\ U_2^T \end{pmatrix} S_b (U_1, U_2) + \begin{pmatrix} U_1^T \\ U_2^T \end{pmatrix} S_w (U_1, U_2) \qquad (6.35)$$

$$= \begin{pmatrix} U_1^T S_b U_1 & U_1^T S_b U_2 \\ U_2^T S_b U_1 & U_2^T S_b U_2 \end{pmatrix} + \begin{pmatrix} U_1^T S_w U_1 & U_1^T S_w U_2 \\ U_2^T S_w U_1 & U_2^T S_w U_2 \end{pmatrix}$$

$$\Rightarrow U^T S_b U = \begin{pmatrix} U_1^T S_b U_1 & 0 \\ 0 & 0 \end{pmatrix} \quad \text{and}$$

$$U^T S_w U = \begin{pmatrix} U_1^T S_w U_1 & 0 \\ 0 & 0 \end{pmatrix}$$

Having solved for the reduced SVD of H_t, it was further demonstrated in [46] that the relationship between the *diagonalized* total scatter matrix and the other two scatter matrices could be redefined as shown in (6.36) to get an *identity matrix* on one side of the equation.

$$\Sigma_t^2 = U_1^T S_b U_1 + U_1^T S_w U_1$$
$$\Rightarrow I_t = \Sigma_t^{-1} U_1^T S_b U_1 \Sigma_t^{-1} + \Sigma_t^{-1} U_1^T S_w U_1 \Sigma_t^{-1} \qquad (6.36)$$

The next step of the algorithm involved a *second* SVD, and this time the *full SVD* is computed on *half* of the *between*-class component in (6.36), allowing for the derivation of the between-class scatter matrix *diagonalization* shown in (6.37). In this equation the full SVD is taken over the value given to B. By substituting the result back in the between-class component of (6.36) it can be shown that the *orthogonal* component Q disappears as its product simply becomes an *identity matrix*, while the diagonal matrix becomes the *square* of itself.

$$B = \Sigma_t^{-1} U_1^T H_b, \quad \text{with SVD} \quad B = P\tilde{\Sigma}Q^T$$
$$\Rightarrow \Sigma_t^{-1} U_1^T H_b = BB^T = P\tilde{\Sigma}Q^T Q\tilde{\Sigma}P^T = P\tilde{\Sigma}^2 P^T = P\Sigma_b P^T \qquad (6.37)$$

As a final step in the simultaneous *diagonalization* of scatter matrices, having discovered the diagonalization of the total and between-class scatter matrices, the *orthogonal* matrix $\boldsymbol{P}$ discovered in (6.37) can be *multiplied* with (6.36) to reveal the diagonalization of all three scatter matrices (6.38); this results in the diagonalization matrix $\boldsymbol{X}$, which simultaneously diagonalizes the three scatter matrices and is demonstrated in (6.39).

$$
\begin{aligned}
\boldsymbol{P}^T \boldsymbol{I}_t \boldsymbol{P} &= \boldsymbol{\Sigma}_b + \boldsymbol{P}^T \boldsymbol{\Sigma}_t^{-1} \boldsymbol{U}_1^T \boldsymbol{S}_w \boldsymbol{U}_1 \boldsymbol{\Sigma}_t^{-1} \boldsymbol{P} \\
&\Rightarrow \boldsymbol{I}_t = \boldsymbol{\Sigma}_b + \boldsymbol{P}^T \boldsymbol{\Sigma}_t^{-1} \boldsymbol{U}_1^T \boldsymbol{S}_w \boldsymbol{U}_1 \boldsymbol{\Sigma}_t^{-1} \boldsymbol{P} \\
&\Rightarrow \boldsymbol{P}^T \boldsymbol{\Sigma}_t^{-1} \boldsymbol{U}_1^T \boldsymbol{S}_w \boldsymbol{U}_1 \boldsymbol{\Sigma}_t^{-1} \boldsymbol{P} = \boldsymbol{I}_t - \boldsymbol{\Sigma}_b = \boldsymbol{\Sigma}_w
\end{aligned} \tag{6.38}
$$

$$
\begin{aligned}
\boldsymbol{X}^T \boldsymbol{S}_b \boldsymbol{X} &= \begin{pmatrix} \boldsymbol{\Sigma}_b & 0 \\ 0 & 0 \end{pmatrix} \equiv \boldsymbol{D}_b, \quad \boldsymbol{X}^T \boldsymbol{S}_w \boldsymbol{X} = \begin{pmatrix} \boldsymbol{\Sigma}_w & 0 \\ 0 & 0 \end{pmatrix} \equiv \boldsymbol{D}_w \\[2mm]
\boldsymbol{X}^T \boldsymbol{S}_t \boldsymbol{X} &= \begin{pmatrix} \boldsymbol{\Sigma}_t & 0 \\ 0 & 0 \end{pmatrix} \equiv \boldsymbol{D}_t, \quad \boldsymbol{X} = \boldsymbol{U} \begin{pmatrix} \boldsymbol{\Sigma}_t^{-1} \boldsymbol{P} & 0 \\ 0 & \boldsymbol{I} \end{pmatrix}
\end{aligned} \tag{6.39}
$$

Finding the solution for the ULDA transformation matrix ($\boldsymbol{G}$) can be simplified when the diagonalization matrix ($\boldsymbol{X}$) is available. In this case, the *optimization criterion* from (6.33) can be reduced to the form shown in (6.40), by first substituting the scatter matrices for their *diagonalized* forms, and then simplifying using the *cyclic matrix trace* property together with the *pseudoinverse* properties of equality. Note that with t equal to the total scatter matrix *rank*, the component $\boldsymbol{G}_2$ does not contribute to the optimization.

$$
\begin{aligned}
\boldsymbol{G}^T \boldsymbol{S}_b \boldsymbol{G} &= \boldsymbol{G}^T \left(\boldsymbol{X}^{-1} \right)^T \left(\boldsymbol{X}^T \boldsymbol{S}_b \boldsymbol{X} \right) \boldsymbol{X}^{-1} \boldsymbol{G} = \tilde{\boldsymbol{G}}^T \boldsymbol{D}_b \tilde{\boldsymbol{G}} \\
\boldsymbol{G}^T \boldsymbol{S}_t \boldsymbol{G} &= \boldsymbol{G}^T \left(\boldsymbol{X}^{-1} \right)^T \left(\boldsymbol{X}^T \boldsymbol{S}_t \boldsymbol{X} \right) \boldsymbol{X}^{-1} \boldsymbol{G} = \tilde{\boldsymbol{G}}^T \boldsymbol{D}_t \tilde{\boldsymbol{G}} \\
\tilde{\boldsymbol{G}} &= \boldsymbol{X}^{-1} \boldsymbol{G} = \begin{pmatrix} \boldsymbol{G}_1 \\ \boldsymbol{G}_2 \end{pmatrix} \quad \text{with} \quad \boldsymbol{G}_1 \in \boldsymbol{R}^{t \times q} \\
\text{and} \quad \boldsymbol{G}_2 &\in \boldsymbol{R}^{(m-t) \times q} \quad \text{for} \quad q = \operatorname{rank}(\boldsymbol{S}_b) \\
&\Rightarrow \boldsymbol{G} = \arg \max_{\boldsymbol{G}} \operatorname{tr}\left(\left(\boldsymbol{G}_1^T \boldsymbol{G}_1 \right)^+ \left(\boldsymbol{G}_1^T \boldsymbol{\Sigma}_b \boldsymbol{G}_1 \right) \right) \\
&= \arg \max_{\boldsymbol{G}} \operatorname{tr}\left(\left(\boldsymbol{G}_1 \boldsymbol{G}_1^+ \right)^T \boldsymbol{\Sigma}_b \left(\boldsymbol{G}_1 \boldsymbol{G}_1^+ \right) \right)
\end{aligned} \tag{6.40}
$$

Under this *new* criterion the transformation matrix can be seen as being formed by the product of $\boldsymbol{X}$ with some *unknown* matrix ($\tilde{\boldsymbol{G}}$). In [46] Ye presented a theorem which demonstrated how the maximization of this criterion need only depend on a *truncated* form of the diagonalization matrix, namely the *submatrix* formed by first q columns in $\boldsymbol{X}$ (the columns representing the *eigenvectors* corresponding to $\boldsymbol{S}_b$'s nonzero *eigenvalues*). In applying this theorem, it was shown that $\tilde{\boldsymbol{G}}$ could be decomposed into two different components of which only *one* would have any effect on the maximization of the criterion, the $t \times q$ dimensional matrix $\boldsymbol{G}_1$ (for the t rank total scatter matrix), and that this component could be decomposed further to a non-singular $q \times q$ matrix $\boldsymbol{M}$ via removing rows found to have no influence over the criterion. Consequently M could be used in the place of $\tilde{G}$ as the unknown criterion-influencing component, and, when combined with the truncated diagonalization matrix X_q, Ye's theorem proved that the resulting *generalized* transformation matrix (6.41) would be guaranteed to maximize the criterion given any *nonsingular* value of M; with X_q in the equation being the first q (as defined in (6.40)) vectors in diagonalization matrix. In the specific instance of ULDA the value of M is set to the *identity matrix* (I_q) so the transformation matrix becomes equivalent to X_q (6.42), and the reduced dimensional space produces features that are *uncorrelated* from one another. Assigning the discovered value of G to the formerly described reduced LDA classifier then gives us our ULDA classifier.

$$G = X\tilde{G} = X_qM, \quad R^{q \times q} \tag{6.41}$$

$$G = X_q \tag{6.42}$$

In [30] Park et al. demonstrated that the relationship linking the scatter matrices to the reduced LDA optimization criterion could be reformulated so as to allow for the maximization of the criterion over a higher dimensional *kernel space*. A follow up on this research was done by Wang et al. [42], who showed that this new kernel-based *optimization criterion* could be used in the place of the non-kernel-based optimization criterion when performing ULDA for feature extraction; the resulting algorithm was termed *Kernel Uncorrelated Discriminant Analysis* or KUDA. In its use for classification, the KUDA algorithm can be derived by simply replacing non-kernelized input with equivalent *kernelized* input (6.43) during the training and classification phases; note that the inputs shown in (6.43) are derived with $\kappa(x, y)$ being the kernel function and with the input data matrix $A \in R^{n \times m}$ of the non-kernelized formulation replaced by its *Gram kernel matrix* K. Moreover, the kernel *transformation* function $\emptyset(x)$ again projects the sample x into the kernel space with respect to the training data, and, to solve for the scatter matrices, a subset of kernelized samples K_i is obtained for each class i belonging to the k different classes. Hence, during the KUDA *training phase* the two *half*-scatter matrices, H_b and H_t, would then be replaced with their kernelized forms, $H_{b(\emptyset)}$ and $H_{t(\emptyset)}$, to solve for the transformation matrix (G); while during the classification phase the *within*-class scatter matrix (S_w) and *input* sample vector (x) would be replaced with their respective kernelized equivalents, $S_{w(\emptyset)}$ and $\emptyset(x)$ to obtain the class conditional probability density functions. For computational efficiency, the KUDA classifier takes advantage of the *kernel trick*, a technique we previously discussed in the formulation of the kernel SVM classifier (Sect. 6.3). The work done by Park et al. [46] showed that all reduced LDA inputs could be reformulated as a grouping of *dot products* and these dot products could be replaced by kernel functions, which, in turn, would still produce dot product values, but this time between a projection of the inputs in a higher dimensional kernel space, moreover without actually needing to perform any computationally expensive feature space projections. Consequently, as was previously demonstrated for the SVM classifier, the samples of a dataset, when treated in the kernel space, may become *separable* where such separation would not otherwise have been possible using the non-kernelized version of the ULDA classifier.

$$K(x_1, x_2 \ldots, x_n) = \begin{bmatrix} \kappa(x_1, x_1) & \cdots & \kappa(x_1, x_n) \\ \vdots & \ddots & \vdots \\ \kappa(x_n, x_1) & \cdots & \kappa(x_n, x_n) \end{bmatrix}, \quad x_i \epsilon A$$

$$\emptyset(x) = [\kappa(x, x_1), \kappa(x, x_2) \ldots \kappa(x, x_n)], \quad x_i \epsilon A$$

$$c_{\emptyset} = \frac{1}{n} K e, \quad e = (1, 1, \ldots, 1)^T \in R^n,$$

$$c_{\emptyset}^{(i)} = \frac{1}{n_i} K_i e^{(i)}, \quad e^{(i)} = (1, 1, \ldots 1)^T \in R^{n_i} \tag{6.43}$$

$$H_{t(\emptyset)} = \frac{1}{\sqrt{n}} \left(K - c_{\emptyset} e^T \right),$$

$$H_{b(\emptyset)} = \frac{1}{\sqrt{n}} \left[\sqrt{n_1} \left(c_{\emptyset}^{(i)} - c_{\emptyset} \right), \ldots, \sqrt{n_k} \left(c_{\emptyset}^{(i)} - c_{\emptyset} \right) \right]$$

$$S_{w(\emptyset)} = \frac{1}{n} \sum_{i=1}^{k} \sum_{x \in K_i} \left(x - c_{\emptyset}^{(i)} \right) \left(x - c_{\emptyset}^{(i)} \right)^T$$

For the purpose of our GRF research, we implemented both the ULDA and KUDA variants of the LDA classifier, using a C# port [38] of the popular *Jama* matrix manipulation package [15] to perform the required matrix decompositions where required. In our implementation, the Jama SVD algorithm was modified so to allow for the calculation of either the *full* or *economy* SVD when requested, as opposed the default Jama SVD behavior that produces only the economy SVD. We also corrected for a *deficiency* in Jama, whereby the SVD algorithm fails when calculated on matrices having more columns than rows. With regard to *parameter optimization*, the ULDA algorithm can run *independent* of external configuration parameters, whereas the optimization of the KUDA algorithm depends on the *kernel function* configuration used to run it. We decided to use the *Gaussian RBF* kernel (6.21) for our KUDA classifier, leaving us with a single kernel parameter (γ) to optimize; optimization was accomplished using an *exhaustive search* of values. Additionally, to remove any potential *bias* due to variations in feature scale, we *rescaled* all input data to these LDA-based classifiers to fall within the range of (0,1); while, using our knowledge of the potential subject–sample input *distribution*, the a priori probabilities ($P(y_i)$ in (6.26)) of all 10 subjects were considered to be equal at 0.1. Finally, unlike the previous classifiers discussed in this chapter, our LDA-based classifiers required no extra work be done to determine *posterior probabilities* and the classifier outputs (6.26) were used *directly* in our EER calculations.

Our initial attempts at running the ULDA and KUDA classifiers against our GRF data achieved relatively poor recognition results. We discovered that these results were caused by two different problems with our chosen implementation. The first of these problems was related to our application of *Bayes' rule* to calculate posterior probabilities. In our initial implementation, we neglected the fact that the *distances* in the exponent of the unscaled class probabilities may take on *very large negative values*, some so large that they resulted in *underflow* in a typical software application. This led to cases where all class probabilities would end up being assigned a value of *zero* and the scaled classifier posterior probability would come out as being *undefined*. To address this issue, we added an additional step prior to applying Bayes' rule. The new step involves computing the *logarithms* for each of the class probabilities, then subtracting a *constant* from each of the logarithms to assign a value of zero to the greatest logarithmic probability, and finally *exponentiating* the logarithmic terms (6.44). In this equation, we calculate the logarithms for all our *class probabilities* (the *conditional* probabilities multiplied with the a priori class probabilities), then subtract a value equivalent to the *maximum* logarithm $C_{max_{y_j}}$ for all k classes. After applying this step, we find that at least one of the *unscaled* class probabilities will always be assigned a value of *one* and the classifier's *posterior probability* output will never be undefined.

$$
\begin{aligned}
L_{y_i} &= \ln(p(x|y_i)) + \ln(p(y_i)), \\
L_i &= L_{y_i} - C_{max_{y_j}}, j = 1, \ldots k, \\
P(y_i|x) &= \frac{e^{L_i}}{\sum_{j=1}^{k} e^{L_j}}
\end{aligned}
\qquad (6.44)
$$

The second problem we encountered in our LDA implementation came as the result of our chosen LDA dimensionality reduction technique. What we discovered was that the ULDA-based dimensionality techniques had a strong tendency to *overfit* our training data. Consequently, our dimensionally reduced training data class distributions, and, by extension our *pooled within-class covariance matrix*, contained an incredibly *small* degree of *variance*, producing class boundaries that *perfectly* separated training samples by class yet proved poor for classification of unseen *testing* data samples. In [47], it was found that the ULDA transformation causes the samples of each class to *converge* to a single point per class under the *mild condition*, which occurs when the *rank* of the sum of the *within-* and *between-*class scatter matrices is equal to that of the total scatter matrix (6.45) and typically leads to classifier *overfitting*. In that paper, the *regularized* LDA technique known as RLDA was suggested as an overfitting-resistant alternative. For our research, we altered the handling of our ULDA output to assume a greater degree of *variance* in the calculation of our class probabilities. Taking into account the fact that the ULDA dimensionality reduction leads to features that are *uncorrelated* (producing diagonal covariance matrices with the variances of each feature represented down the diagonal) we replaced the transformed *within*-class scatter matrix of (6.31) with a number of different diagonal matrices and tested the classifier performance against them. In our work we found that by simply using an *identity matrix* in the place of the much smaller transformed within-class scatter matrix, the classifier was better

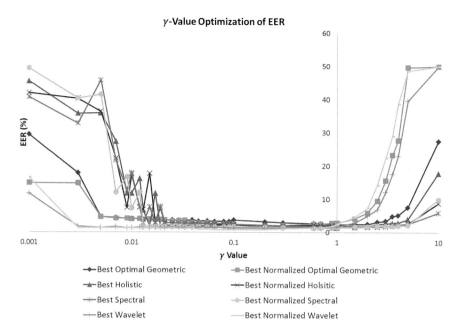

Fig. 6.11 KUDA parameter optimization. This figure demonstrates the impact of the kernel optimized parameter γ on the cross-validated EER calculated across our best performing feature preprocessing techniques

able to generalize class boundaries, providing far better classification results. The new *class conditional probability* function used in our work is shown in (6.46), while the optimization of the *kernel parameter* for the KUDA algorithm under this new classification approach is demonstrated in Fig. 6.11. In this variant of the class conditional probability definition, the *diagonal covariance matrix* (E) is determined based on its ability to *generalize* class boundaries. In our *implementation*, as mentioned, the *identity matrix* was chosen for this q-dimensional space, simplifying the calculations required.

$$\mathrm{rank}(S_b) + \mathrm{rank}(S_w) = \mathrm{rank}(S_t) \tag{6.45}$$

$$
\begin{aligned}
P(x||y_i) &\approx \frac{1}{(2\pi)^{\frac{m}{2}}|E|^{\frac{1}{2}}} \exp\left(\frac{1}{2}\left(x - \mu_{y_i}\right)E^{-1}\left(x - \mu_{y_i}\right)\right) \\
\Rightarrow P(x|y_i) &\approx \frac{1}{(2\pi)^{\frac{m}{2}}} \exp\left(\frac{1}{2}\left(x - \mu_{y_i}\right)\left(x - \mu_{y_i}\right)\right),
\end{aligned}
\tag{6.46}
$$
$$\text{when} \quad E = I_q$$

Table 6.4 LDA classifier performance

Optimal ULDA classifier results			
Preprocessor	Threshold	CV EER (%)	EER incr. (%)
Best optimal geometric	0.0468	7.12222	−434.1
Best normalized optimal geometric	0.0812	5.45555	−2968.8
Best holistic	0.1937	1.5	41.3
Best normalized holistic	0.2546	1.5	26.6
Best spectral	0.3046	1.21111	40.1
Best normalized spectral	0.2796	1.16666	36.7
Best wavelet	0.0625	3.81111	−195.6
Best normalized wavelet	0.0812	2.65555	−141.4

Optimal KUDA classifier results				
Preprocessor	γ	Threshold	CV EER (%)	EER incr. (%)
Best optimal geometric	1	0.1171	2.3	−72.5
Best normalized optimal geometric	0.2	0.1734	1.87777	−956.2
Best holistic	0.9	0.1671	1.45555	43
Best normalized holistic	2.5	0.1531	1.75555	14.1
Best spectral	1.5	0.2015	1.52222	24.7
Best normalized spectral	0.6	0.1656	1.26666	31.3
Best wavelet	0.09	0.3156	1.24444	3.4
Best normalized wavelet	0.1	0.289	0.78888	28.2

These tables demonstrate the best performance achieved by the ULDA and KUDA classifiers for each preprocessing technique. The *threshold* shown is the *posterior probability* threshold at which the EER improvement was calculated and the EER improvement represents the *improvement* in recognition performance achieved by the optimal LDA variant over the results calculated in the previous two chapters

The optimal GRF recognition results for our implementation of the ULDA and KUDA classifiers over our best previously discovered feature spaces are demonstrated in Table 6.4. We found that the *kernelized* ULDA classifier performed considerably better than the ULDA classifier on the wavelet and geometric feature spaces, but was generally not as strong at classifying the holistic and spectral feature spaces. Moreover, we found that these LDA classifier variants generally performed slightly worse than SVM classifier over the holistic and spectral feature spaces, yet performed noticeably better than any other in the wavelet feature spaces. Also, as was the case in our previous SVM and MLP classifiers, we found that our LDA variances performed worse than KNN in the geometric feature spaces, but, again, this may be explained by the *training bias* toward KNN. Our findings paralleled those of [27] in that our LDA classifier achieved similar results to those of the SVM classifier; however, little detail was given in [27] regarding the LDA *variant* used, and, to the best of our knowledge, we are the first to use ULDA and KUDA for GRF recognition. Additionally, it must be noted that these classification results were idealized in that real a priori probabilities were actually known during computations, something that is often not the case in other classification scenarios.

Nevertheless, our findings leave a lot of room for future improvement; for instance, classification performance could be improved by deriving *class conditional covariance* matrices (6.46) that better reflect the *variance* of features in the reduced LDA *space* than our chosen *identity matrix*, or, alternatively, the powerful dimensionality reduction abilities of these techniques could be repurposed to discover more discriminative features during *feature extraction*.

6.5   Least Square Probabilistic Classifier

The results achieved by the LDA classifier described in the previous section demonstrated that *generative* classifiers can in fact be effective for the purpose of GRF recognition. The LDA classifier, however, has several *drawbacks*, including the fact that the previously discussed LDA algorithms required the use of intensive computations, such as SVD, on matrices that grow with the size of the training dataset, as can be seen from the derivation of the *half-total scatter matrix* in (6.34). This problem could potentially be mitigated by using LDA as a *binary classifier* and performing classification through the *one-against-one* strategy previously used with the SVM classifier, but this would take away from the convenience of having a classifier that directly generates single dataset-wide probabilistic output values and might reduce recognition performance. An alternative classification technique proposed by Sugiyama in [40] was designed to directly model *posterior probability*, but in a manner that would not require the performing of computations on any matrix with a dimensionality larger than the *largest subset of class samples* in the training data. This efficient algorithm, referred to as *Least Squares Probabilistic Classification* (LSPC), opts to solve for optimal posterior probability models via the *Least Squares* learning technique, as opposed to relying on the multivariate Gaussian modeling that forms the core of the posterior probability derivation in the LDA classification approach. To our knowledge the LSPC classification technique has never before been used for the purpose of GRF recognition, however, in his paper Sugiyama demonstrated that the algorithm could achieve strong classification performance and fast training times for complicated datasets including handwritten digits and satellite imagery.

The LSPC classifier is a *discriminative eager* learning-based classification algorithm that generates its *probabilistic models* by first deriving the class probability *search spaces* as a *parameterized* linear combination of training data-based *basis functions*, and then solves for this system of *linear equations* to discover the probability models that most closely reflect the *true posterior class probability* spaces. In LSPC, the individual parameterized *class models* take the form demonstrated in (6.47), where the parameter α is optimized to correspond with the value it would take for the optimal probability representation of the training data. This probability model reflects the *likelihood* that a given sample x belongs to a given class y and is composed of a series of b basis functions (ϕ) corresponding to a sequence of sample pairs, each producing a value greater than *zero*.

$$q(y|x;\alpha) = \sum_{l=1}^{b} \alpha_l \emptyset_l(x,y) = \alpha^T \emptyset(x,y),$$

$$\text{where} \quad \emptyset(x,y) \geq \mathbf{0}_b, \quad \forall(x,y) \tag{6.47}$$

Having established the parameterized posterior class probability models, the LSPC *optimization criterion* can be represented as the *mean squared error* between the aforementioned parameterized probability models and the *true probability values*. This formulation of the optimization criterion for the LSPC classifier is presented in (6.48), where it can be seen that the criterion is represented as the mean squared error between the *true probability* $p(y|x)$ and *modeled probability* $q(y|x;\alpha)$ across the samples from all the training classes (c) for some constant B. In this case, the *true probability* values can be established via analysis of the training data as per (6.49), which demonstrates that the unknown probability density components that form the values for H and h in the optimization criterion can be *estimated* by computing the *sample averages*.

$$(\alpha) = \frac{1}{2}\sum_{y=1}^{c}\int (q(y|x;\alpha) - p(y|x))^2 p(x)dx$$

$$= \frac{1}{2}\sum_{y=1}^{c}\int q(y|x;\alpha)^2 p(x)dx - \frac{1}{2}\sum_{y=1}^{c}\int q(y|x;\alpha)p(x,y)dx + B \tag{6.48}$$

$$= \frac{1}{2}\alpha^2 H\alpha - h^T\alpha + B$$

$$H = \sum_{y=1}^{c}\int \emptyset(x,y)\emptyset(x,y)^T p(x)dx$$

$$\approx \hat{H} = \frac{1}{n}\sum_{y=1}^{c}\sum_{i=1}^{n}\emptyset(x,y)\emptyset(x,y)^T, \tag{6.49}$$

$$h = \sum_{y=1}^{c}\int \emptyset(x,y)p(x,y)dx \approx \hat{h} = \frac{1}{n}\sum_{i=1}^{n}\emptyset(x_i,y_i)$$

The equation resulting from (6.48) is *convex* and thus finding its *global minimum*, which occurs when its derivative is equal to zero, will reveal the value of α corresponding to the *optimal probability model* ($\hat{\alpha}$), as can be seen in (6.50). To avoid the potential for model *overfitting* the solution to α contains an additional *L2*

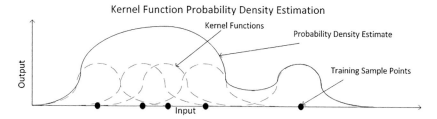

Fig. 6.12 Kernel function probability density estimate. This figure demonstrates how localized kernels could be combined to give the highest probability density estimate values in regions with many samples and lower values elsewhere

regularization term [10], where the regularization input parameter λ is a scalar value that must be determined prior to training.

$$
\hat{\alpha} = \underset{\alpha \in R^b}{\mathrm{argmin}} \left[\frac{1}{2} \alpha^2 H \alpha - \hat{h}^T \alpha + \lambda \alpha^T \alpha \right]
$$

$$
\Rightarrow \left(\hat{H} + \lambda I_b \right) \alpha = \hat{h} \quad \Rightarrow \hat{\alpha} = \left(\hat{H} + \lambda I_b \right)^{-1} \hat{h}
$$

(6.50)

In [40] Sugiyama demonstrated that a final posterior class probability could be acquired by taking the aforementioned probability models, rounding any *negative outputs* to *zero*, and then performing a *normalization step*; yet he went on to show that the solution could be found more efficiently when appropriate *basis functions* are chosen. Sugiyama chose to separate the basis input and output parameters (*x* and *y*) so that each was transformed by a different kernel function, with the *Kronecker delta* kernel [19] chosen to handle the output values. Having formed the *basis* from these two different kernel functions, the parameterized probability models can take the form shown in (6.51). In this equation, the basis function from (6.47) is split into two different kernel functions, with the sample input values (*x*) forming the parameters for some arbitrary kernel function and the sample output class labels (*y*) forming the *parameters* for the Kronecker delta function. In this case, *n* represents the number of samples in the training dataset and $K(x, x_l)$ the arbitrary kernel function. Alternatively, accounting for the *effect* of the *delta function*, the models could be computed separately in a *class-wise* manner as demonstrated in (6.52).

$$
q(y|x; \alpha) = \sum_{y'=1}^{c} \sum_{l=1}^{n} \alpha_l^{(y')} K(x, x_l) \delta_{y,y'},
$$

$$
\delta_{y,y'} =
\begin{cases}
1 & \text{if } y = y' \\
0 & \text{otherwise}
\end{cases}
$$

(6.51)

$$q(y|\boldsymbol{x}; \boldsymbol{\alpha}) = \sum_{l=1}^{n_y} \alpha_l^{(y)} K\left(\boldsymbol{x}, \boldsymbol{x}_l^{(y)}\right) \qquad (6.52)$$

Using the above discussion, Sugiyama showed that further simplification could be accomplished by choosing a *localized* kernel to handle the basis input parameter; a localized kernel being a kernel whose values are at their greatest nearest *known class* contributing training points and become smaller as you move further away from those points. In this case the kernels would make the greatest contributions to probability values in regions where kernels *overlap* and little-to-no contribution in regions with few or no training samples for a given class (see Fig. 6.12). Consequently, when using a localized kernel, such as the *Gaussian kernel* (the kernel used in the SVM and LDA classifiers), the class probability models could be *reduced* to the form shown in (6.53). In this case the kernel function need only be computed over the n_y samples belonging to the examined class y.

$$q(y|\boldsymbol{x}; \boldsymbol{\alpha}) = \sum_{l=1}^{n_y} \alpha_l^{(y)} K\left(\boldsymbol{x}, \boldsymbol{x}_l^{(y)}\right) \qquad (6.53)$$

In making the *simplification* of (6.53), the maximum dimensionality of the matrix $\boldsymbol{H}$ would be reduced from a square matrix with a dimensionality equivalent to the number of samples in the training dataset (n) to a square matrix with a dimensionality only as large as the number of training samples for the *examined* class y (n_y); this drastically decreases the *computational work* needed to solve for the *optimal* value of $\boldsymbol{\alpha}$, as shown in (6.54). In this case $\boldsymbol{\alpha}^{(y)}$ must be calculated for each class y, but now the matrix ($\boldsymbol{H}$) and vector ($\boldsymbol{h}$) required for the optimization will only ever have a dimensionality equal to the number of training samples in the class being optimized. With this more efficient approach to computing the probability models, the final *posterior class probabilities* could be computed using the Sugiyama's [40] *negative value rounding* and *normalization* approach (6.55), where the class probability model is normalized by the sum of all other parameterized class probability models and any *negative estimates* produced by the $\boldsymbol{\alpha}$ parameter models are *rounded* up to *zero*.

$$\hat{H}_{l,l'}^{(y)} = \frac{1}{n}\sum_{i=1}^{n} K\left(\boldsymbol{x}_i, \boldsymbol{x}_l^{(y)}\right) K\left(\boldsymbol{x}_i, \boldsymbol{x}_{l'}^{(y)}\right),$$

$$\hat{h}_l^{(y)} = \frac{1}{n}\sum_{i=1}^{n_y} K\left(\boldsymbol{x}_i^{(y)}, \boldsymbol{x}_l^{(y)}\right), \qquad (6.54)$$

$$\boldsymbol{\alpha}^{(y)} = \left(\alpha_1^{(y)}, \dots, \alpha_{n_y}^{(y)}\right)^T, \quad \boldsymbol{\alpha}^{(y)} = \left(\hat{H}^{(y)} + \lambda \boldsymbol{I}_{n_y}\right)^{-1} \hat{h}^{(y)}$$

$$\hat{p}(y|x) = \frac{\max\left(0, \ \sum_{l=1}^{n_y} \hat{a}_l^{(y)} K\left(x, x_l^{(y)}\right)\right)}{\sum_{y'=1}^{c} \max\left(0, \sum_{l=1}^{n_{y'}} \hat{a}_l^{(y')} K\left(x, x_l^{(y')}\right)\right)} \tag{6.55}$$

For the purpose of our research we initially implemented the efficient variant of the LSPC algorithm described above, making use of the C# port [38] for the *Jama* matrix manipulation package to perform the α optimization step. Externally, this left us with *three configurable inputs* to be accounted for when performing classification: the *input* training feature values, the *regularization* parameter, and the *kernel* parameters. To mitigate any potential *bias* in the feature input values, we went with the *rescaling* technique used in our previous four classifiers and rescaled the input for each feature input to fall between the values of 0 and 1 prior to training and classification. The remaining input parameters were assigned by the LSPC classification algorithm to take some *predetermined* performance-optimizing values and thus required some *tuning* to achieve desirable performance. Moreover, the LSPC classifier was designed such that it could be assigned *any* arbitrary localized kernel function. In our implementation, we used the variant of *Gaussian kernel* previously used in the implementation by Sugiyama in [40] (6.56), leaving us with a single kernel parameter σ. This allowed us to optimize our GRF recognition performance by performing an exhaustive 2-dimensional grid search over the regularization and kernel parameters using an analytically determined set of possible values for each.

$$K\left(x, x'\right) = \exp\left(-\frac{x - x'^2}{2\sigma^2}\right) \tag{6.56}$$

During our initial optimization analysis of the LSPC algorithm we encountered several *problems* regarding numerical edge cases and computational limitations. First of all, the original LSPC algorithm implicitly assumed that for any given input value at least one of the posterior class probability models will produce a value *greater than zero*. In practice, using our GRF training data, this assumption proved to be *untrue* and, as a result, *undefined* posterior probabilities were encountered. Furthermore, we found that the *kernel exponent* terms in the LSPC classifier function (6.55) took on very *large negative values* to the point that we frequently encountered arithmetic *underflow* during our GRF recognition tests and were unable to assign *nonzero* posterior class probabilities. To address these issues we made several modifications to the original algorithm. With regards to the assumption of there being at least one *nonzero* posterior class probability we first calculated the sum of all class probability model outputs, and in the case of a zero value probability sum we assigned a value of *zero* for the *requested* posterior class probability. Thus in the case where only negative model probability outputs were encountered *none* of the trained GRF subjects (classes) would be accepted as the *owner* of the given input sample. To avoid the possibility of *arithmetic underflow*, we used a trick similar to the *logarithmic trick* for avoiding arithmetic underflow discussed in Sect. 6.4. In this case, we adjusted the algorithm to search for an exponent by which to divide our *kernelized* samples to give our largest exponent term, corresponding to a positive α, a value of 1, while avoiding underflow by applying the *law of exponents* for merging division terms. To account for the fact that in some cases the largest positive exponential term corresponded with *probability models* that came out to be *negative* overall, we modified our search for exponent denominator to *ignore* models with negative outputs. Our *new* posterior class probability output function resulting from these changes is demonstrated in (6.57); in this equation probability estimates are assumed to have a value of *zero* when the sum of the probabilities found across known classes is equal to zero and exponential terms are adjusted by the value E_{max} to avoid arithmetic underflow. Using our modified implementation of the LSPC classifier, the best achieved GRF recognition rates for each tested parameter value are demonstrated in Fig. 6.13.

$$E_{\max} = \max_{q(y|x;\,\boldsymbol{\alpha}) > 0,\, \boldsymbol{\alpha}_l^{(y)} > 0} \left(-\frac{\left\| x - x_l^{(y')} \right\|^2}{2\sigma^2} \right),$$

$$L(y|x) = \frac{1}{e^{E_{\max}}} \frac{\max\left(0,\ \sum_{l=1}^{n_y} \hat{\alpha}_l^{(y)} \exp\left(-\frac{\|x-x_l^{(y)}\|^2}{2\sigma^2} \right) \right)}{\sum_{y'}^{c} \max\left(0,\ \sum_{l=1}^{n_{y'}} \hat{\alpha}_l^{(y')} \exp\left(-\frac{\left\| x-x_l^{(y')} \right\|^2}{2\sigma^2} \right) \right)} \tag{6.57}$$

$$= \frac{\max\left(0, \sum_{l=1}^{n_y} \hat{\alpha}_l^{(y)} \exp\left(-\frac{\|x-x_l^{(y)}\|^2}{2\sigma^2} - E_{\max} \right) \right)}{\sum_{y'}^{c} \max\left(0,\ \sum_{l=1}^{n_{y'}} \hat{\alpha}_l^{(y')} \exp\left(-\frac{\left\| x-x_l^{(y')} \right\|^2}{2\sigma^2} - E_{\max} \right) \right)},$$

$$\hat{p}(y|x) = \begin{cases} L(y|x) & \text{for } L(y|x) > 0 \\ 0 & \text{for } L(y|x) < 0 \end{cases}$$

After running our results, we found that the LSPC parameter optimization curves for our *kernel* (σ) and *regularization* (λ) parameters mirrored our observed findings for the performance of equivalent parameter optimizations with the SVM classifier (Sect. 6.3); it can be seen that GRF recognition performance tended to be better for both smaller kernel parameter values (less than 1) and larger regularization parameter values (greater than 1). With regard to overall GRF recognition performance (Table 6.5), the cross-validated results that the LSPC classifier achieved were better than all other non-KNN classifiers on the geometric feature spaces,

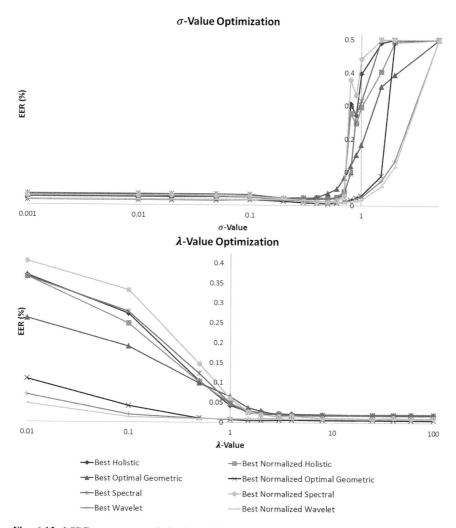

Fig. 6.13 LSPC parameter optimization. This figure compares our best cross-validated EER values achieved with the LSPC classifier to the optimization parameters used to achieve them for all our best performing preprocessing techniques. For both given parameter values optimization was carried out by testing the given parameter with every possible value for the other parameter and returning the EER for the best combination

while it also consistently performed better than the KNN classifier on all nongeometric feature spaces. Of particular note was the strong recognition performance by the LSPC classifier when applied to the *wavelet* feature spaces, with the LSPC classifier achieving the best performance of all classifiers in the non-normalized wavelet feature space. This together with its performance for the geometric and LLSRDTW-normalized spectral feature spaces may suggest that the LSPC classifier *benefitted* more than other classifiers when applied to feature spaces that were

Table 6.5 LSPC classifier performance

Optimal LSPC classifier results

Preprocessor	λ	σ	Thres.	CV EER (%)	EER incr. (%)
Best optimal geometric	8	0.3	0.1921	1.8	−35
Best normalized optimal geometric	1.5	0.5	0.314	0.51111	−187.5
Best holistic	25	0.4	0.1581	1.92222	24.7
Best normalized holistic	40	0.2	0.2453	1.77777	13
Best spectral	4	0.4	0.1822	1.82222	9.8
Best normalized spectral	3	0.6	0.1453	1.06666	42.1
Best wavelet	1	0.6	0.3703	0.94444	26.7
Best normalized wavelet	1	0.6	0.3187	0.87777	20.2

This table demonstrates the best performance achieved by the LSPC classifier for each preprocessing technique. The *threshold* shown is the threshold at which the EER value was calculated (a value between 0 and 1 derived from the *raw posterior probability output*) and the EER *improvement* represents the improvement in recognition performance achieved by the optimal LSPC variant over the results calculated in the previous two chapters

Table 6.6 Classifier performance comparison

Classifier GRF recognition performance

Feature space	Normalizer	Best classifier	EER (%)
Optimal geometric	–	KNN	1.33333
Optimal geometric	LLSR	KNN	0.17777
Holistic	–	SVM	0.96666
Holistic	L1	SVM	1.38888
Spectral	–	LDA (ULDA)	1.21111
Spectral	LLSRDTW	SVM	0.83333
Wavelet	–	LSPC	0.94444
Wavelet	LTN	LDA (KUDA)	0.78888

This table compares the best GRF recognition performance achieved across each preprocessor when used in combination with our five chosen classification techniques

extracted using greater degrees of *supervision* during extractor training. Nevertheless, the GRF recognition results achieved by the LSPC classifier in this section have demonstrated that it can be competitive with other classifiers without the large training dataset size-based performance penalty that would be experienced in many other algorithms that *model* posterior class probabilities such as the previously examined LDA.

6.6 Summary

This chapter covered the use of classifiers for the purpose of performing footstep GRF-based person recognition. In the gait biometric-based system previously discussed in Chap. 2, the classification component is the part responsible for learning patterns from the data and producing an output value that can be used to either accept or reject a data sample as belonging to a data class (in verification mode), or identify the class that most closely matches a sample (in identification mode). For the purpose of our research all results were collected in verification mode and returned in the form of an EER. Consequently, each classifier examined in this chapter was configured to accept a sample and a GRF subject label as input then output a single posterior probability representing the classifier's confidence that the provided sample belonged to the provided subject. Through our introduction to the concept of classification, we discussed how classifiers could be categorized as being either instance-based or eager learning-based in addition to following either a discriminative or generative model, with examples of each category of classifier being examined. Five different powerful classification techniques were studied, four of which were examined in previous GRF-based recognition studies, and one (LSPC) that had never before been tested for the purpose of GRF-based recognition.

A theoretical background was provided as each of our selected classifiers were examined, exposing areas in which each could benefit from the optimization of internal parameters. The best results obtained after performing these optimizations over each classifier, in combination with our top performing preprocessors from Chaps. 4 and 5, are demonstrated in Table 6.6. These results may appear to indicate that the normalized optimal geometric preprocessor, in combination with the KNN classifier, performed best; however, as was mentioned in the proceeding sections, the results obtained from our development dataset were strongly biased in favor of the KNN classifier, on account of its use in optimizing the preprocessors. In the chapters that follow, we take the top performing biometric system configurations from the design phase of our demonstrative experiment, those discovered through this and the previous two chapters, and apply them to a set of previously unseen data (our evaluation dataset) to perform the evaluation phase of our demonstrative experiment. Having limited the bias toward any particular classifier or preprocessor, the aim of the next experiment phase is to gather the results needed to better assess our GRF recognition performance and verify the normalization and shoe type assertions introduced in our first chapter objectives.

References

1. Agapitos, Alexandros, O'Neill Michael, and Anthony Brabazon. 2011. Maximum margin decision surfaces increased generalisation in evolutionary decision tree learning. In *Genetic programming: 14th european conference*, 61–72, Torino.
2. Balachander, Thiagarajan, Ravi Kothari, and Hernani Cualing. 1997. An empirical comparison of dimensionality reduction techniques for pattern classification. In *Artificial neural networks—7th international conference*, 589–594, Lausanne.
3. Bottou, Léon, and Chih-Jen Lin. 2007. Support vector machine solvers. In *Large-scale kernel machines*. Bottou Léon et al., eds. chap. 1. 1–28. Cambridge, MA, USA: MIT Press.
4. Christopher J. C. Burges, 1998. A tutorial on support vector machines for pattern recognition. *Data Mining and Knowledge Discovery* 2(2): 121–167.
5. Cattin, Philippe C. 2002. Biometric authentication system using human gait. Ph.D Thesis. Zurich, Switzerland: Swiss Federal Institute of Technology.
6. Chang, Chih-Chung, and Chih-Jen Lin. 2001. LIBSVM: A library for support vector machines. Technical Report. Taipei: National Taiwan University, http://www.csie.ntu.edu.tw/~cjlin/papers/libsvm.pdf.
7. Chen, Pai-Hsuen, Chih-Jen Lin, and Bernhard Schölkopf. 2005. A tutorial on v-support vector machines. *Appl Stochastic Models Bus Ind* 21: 111–136.
8. Fernández-Redondo, Mercedes, and Carlos Hernández-Espinosa. 2001. Weight initialization methods for multilayer feedforward. In *European symposium on artificial neural networks*. 119–124. Bruges.
9. Friedman, Jerome H, 1989. Regularized discriminant analysis. *Journal of the American Statistical Association* 84(405): 165–175.
10. Hachiya, Hirotaka, Masashi Sugiyama, and Naonori Ueda. 2012. Importance-weighted least squares probabilistic classifier for covariate shift adaptation with application to human activity recognition. *Neurocomputing* 80: 93–101.
11. Hajek, Milan. 2005. Models of a neuron. In *Neural networks*. Durban, South Africa: University of KwaZulu-Natal, 9–10.
12. Hastie, Trevor, Andreas Buja, and Robert Tibshirani. 1995. Penalized discriminant analysis. *The Annals of Statistics* 23(1): 73–102.
13. Hastie, Trevor, Robert Tibshirani, and Andreas Buja. 1994. Flexible discriminant analysis by optimal scoring. *Journal of the American Statistical Association* 89(428): 1255–1270.
14. Heaton Jeff. 2014. Encog machine learning framework. http://www.heatonresearch.com/encog.
15. Hicklin Joe et al. 2012. Jama: A java matrix package. http://math.nist.gov/javanumerics/jama/.
16. Hidayat Erwin, Nur A. Fajrian, Azah Kamilah Muda, Choo Yun Huoy, and Sabrina Ahmad. 2011. A comparative study of feature extraction using PCA and LDA for face recognition. In *7th international conference on information assurance and security (IAS)*. 354–359. Melaka.
17. Huang Rui, Qingshan Liu, Hanqing Lu, and Songde Ma. 2002. Solving the small sample size problem of LDA. In *16th international conference on pattern recognition*. 29–32. Quebec City.
18. Keller James M., Michael R. Gray, and James A. Givens, Jr. 1985. A fuzzy K-nearest neighbor algorithm. In *IEEE transactions on systems, man, and cybernetics*. SMC-15(4) 580–585.
19. Kozen Dexter and Marc Timme. 2007. Idefinite summation and the Kronecker delta. http://dspace.library.cornell.edu/bitstream/1813/8352/2/Kronecker.pdf.
20. Lawrence Steve, C. Lee Giles, and Ah Chung Tsoi. 1997. Lessons in neural network training: overfitting may be harder than expected. In *Fourteenth national conference on artificial intelligence*. 540–545. Manlo Park.
21. LeCun Yann, and Yoshua Bengio. 1995. Pattern recognition. In *The handbook of brain theory and neural networks*, Michael A. Arbib, ed.: A bradford book. 864–868.
22. Lin, Hsuan-Tien, Chih-Jen Lin, and Ruby C. Weng. 2007. A note on platt's probabilistic outputs for support vector machines. *Machine Learning* 68(3): 267–276.

23. Liu, Ke, Yong-Qing Cheng, and Jing-Yu. Yang. 1992. A generalized optimal set of discriminant vectors. *Pattern Recognition* 25(7): 731–739.
24. Liu Wei, Yunhong Wang, Stan Z. Li, and Tieniu Tan. 2004. Null space approach of fisher discriminant analysis for face recognition. In *European conference on computer vision, biometric authentication workshop*. 32–44, Prague.
25. Milgram, Jonathan, Mohamed Cheriet, and Robert Sabourin. 2006. One against one or one against all: which one is better for handwriting recognition with SVMs?. In *Tenth international workshop on frontiers in handwriting recognition*, La Baule.
26. Mostayed Ahmed, Sikyung Kim, Mohammad Mynuddin Gani Mazumder, and Se Jin Park. 2008. Foot step based person identification using histogram similarity and wavelet decomposition. In *International conference on information security and assurance*. 307–311. Busan.
27. Moustakidis, Serafeim P., John B. Theocharis, and Giannis Giakas. 2008. Subject recognition based on ground reaction force measurements of gait signals. *IEEE Transactions on Systems, Man, and Cybernetics-Part B: Cybernetics* 38(6): 1476–1485.
28. Orfanidis Sophocles J. 2007. SVD, PCA, KLT, CCA, and All That.
29. Orr Robert J., and Gregory D. Abowd. 2000. The smart floor: A mechanism for natural user identification and tracking. In *CHI '00 conference on human factors in computer systems*. 275–276. The Hague.
30. Park, Cheong Hee, and Haesun Park. 2005. Nonlinear discriminant analysis using kernel functions and the generalized singular value decomposition. *SIAM Journal on Matrix Analysis and Applications* 27(1): 87–102.
31. Phyu, Thair Nu. 2009. Survey of classification techniques in data mining. In *International multiconference of engineers and computer scientists. I IMECS 2009.* 727–731. Hong Kong.
32. Principe, Jose C., Neil R. Euliano, and W. Curt Lefebvre. 1999. Multilayer perceptron. In *Neural and adaptive systems: Fundamentals through simulation*. New York, United States: Wiley, ch. 3, 100–172.
33. Rodríguez, Rubén Vera, Nicholas W. D. Evans, Richard P. Lewis, Benoit Fauve, and John S. D. Mason. 2007. An experimental study on the feasibility of footsteps as a biometric. In *15th European signal processing conference (EUSIPCO 2007)*. 748–752. Poznan.
34. Rodríguez, Rubén Vera, John S.D. Mason, and Nicholas W.D. Evans. 2008. Footstep recognition for a smart home environment. *International Journal of Smart Home* 2(2): 95–110.
35. Rojas, Raúl. 1996. The backpropagation algorithm. In *Neural networks*. ch. 7. 149–180. Berlin, Germany: Springer.
36. Rojas, Raúl. 1996. The biological paradigm. In *Neural networks*. ch. 1. 3–26. Berlin, Germany: Springer.
37. Rynkiewicz, Joseph, 2012. General bound of overfitting for MLP regression models. *Neurocomputing* 90: 106–110.
38. Selormey, Paul. 2004. DotNetMatrix: Simple matrix library for.NET. http://www.codeproject.com/Articles/5835/DotNetMatrix-Simple-Matrix-Library-for-NET.
39. Srivastava, Santosh, Maya R. Gupta, and Béla A. Frigyik. 2007. Bayesian quadratic discriminant analysis. *Journal of Machine Learning Research* 8: 1277–1305.
40. Sugiyama, Masashi. 2010. Superfast-trainable multi-class probabilistic classifier by least-squares posterior fitting. *IEICE Transaction on Information and Systems*. E93-D : 10 2690–2701.
41. Suutala, Jaakko, and Juha Röning. 2008. Methods for person identification on a pressure-sensitive floor: Experiments with multiple classifiers and reject option. *Information Fusion Journal, Special Issue on Applications of Ensemble Methods 9.* 9(1) 21–40.
42. Wang, Ling, Liefeng Bo, and Licheng Jiao. 2006. Kernel uncorrelated discriminant analysis for radar target recognition. In *13th international conference on neural information processing*. 404–411. Hong Kong.
43. Wettschereck, Dietrich. 1994. A study of distance-based machine learning algorithms, Ph.D Thesis. Oregon State University, Corvallis, OR, USA.
44. What is the generalized inverse of a matrix?. http://artsci.wustl.edu/~jgill/papers/ginv.pdf.

45. Wu, Ting-Fan, Chih-Jen Lin, and Ruby C. Weng. 2004. Probability estimates for multi-class classification by pairwise coupling. *Journal of Machine Learning Research* 5: 975–1005
46. Ye, Jieping, 2005. Characterization of a family of algorithms for generalized discriminant analysis on under sampled problems. *Journal of Machine Learning Research* 6: 483–502.
47. Ye, Jieping, et al. 2006. Efficient model selection for regularized linear discriminant analysis. In *15th ACM international conference on information and knowledge management*. 532–539. New York.

Chapter 7
Experimental Design and Dataset

The biometric recognition techniques discussed in the previous three chapters established a set of *components,* which when combined, form what was shown in Chap. 2 to represent a *biometric system.* Before any gait biometric recognition system can be put into practical use, it should first be put through a *design* and *evaluation* process to ensure it has been optimized for the problem at hand and accounts for factors that may potentially affect the performance. Through the work in the previous chapters, we have shown how each particular biometric system component can be optimized to perform GRF-based gait recognition, but in some cases the optimization process can lead to undesirable *bias.* In this chapter, we describe the ways in which gait biometric systems may be evaluated, and then lay out the design for the *evaluation phase* of our demonstrative experiment. This sets us up for an experimental evaluation of the GRF-based gait recognition in the two chapters that follow, with the objective of providing a comprehensive comparison of recognition techniques used in previous research while also demonstrating our own novel methods to increase recognition performance and counter the effects of variations in shoe type or differences in stepping speed.

7.1 Evaluation of Gait Biometric-Based Systems

In El-Abed et al. [3] it was shown that biometric systems can be evaluated according to three criteria: *data quality*, *usability*, and *security* (Fig. 7.1). With respect to the first of these, *data quality*, the aforementioned study pointed to the ISO [6] recommendations for data quality assessment. Following these recommendations, data quality can be measured in terms of *character*, the inherent depth of features and traits that form a biometric sample, *fidelity*, the similarity of a sample to its source, and *utility*, a function of character and fidelity reflecting the impact of a sample on biometric system performance. Applying these data quality measures to the gait biometric we would expect to find a richness of features giving the gait strength in terms of the character metric; however, as we saw in Chap. 2, there is no single piece of equipment that can capture the full set of complex dynamics that form gait and gait analysis is highly influenced by the choice of analysis *approach* selected

Fig. 7.1 Gait biometric system evaluation criteria. The evaluation of a gait biometric system can be approached using one or more of the *criteria* presented in this figure, which is based on the research presented by El-Abed et al. [3]

for its measure. Consequently, the fidelity of the gait biometric might vary considerable depending on analysis approach and its utility may be weakened by our inability to capture the entirety of its character. For *video*-based approaches, samples could face numerous obscuring factors increasing the potential for poor sample fidelity, while *sensor*-based approaches with closer contact to the sensor subject may provide greater fidelity but capture a lesser proportion of the gait character; data utility for a gait biometric approach could then be considered reflective of the *tradeoff* between the fidelity and character for the chosen measurement approach.

An alternative to the challenge of quantifying the strength of a chosen gait biometric approach in terms of data quality would be to instead evaluate the gait biometric based on the second criteria described in El-Abed et al. [3], *usability*. Usability, as defined by the ISO [5] is measured in terms of *efficiency, effectiveness,* and *user satisfaction* with respect to the evaluated biometric system. To measure its *efficiency*, a gait recognition system could be timed at points of user interaction; it then could be evaluated against other biometric systems based on the time used for enrollment, identification, or verification. However, the measure of *effectiveness* is the more commonly applied and arguably the most important form of biometric system evaluation. This metric is reflected by a grouping of success and/or error *metrics* that measure the biometric system's *accuracy* and/or *utilization*. In Chap. 2, we were introduced to several of these metrics including the *false acceptance rate* (FAR) and *false rejection rate* (FRR) for evaluating the biometric system verification; in more practical settings these metrics could be *combined* with another error metric, the *failure to acquire* (FTA) rate, to account for cases of biometric sensor failures, sensor avoidance, and any other circumstances that do not lead to a successful sample capture. In this case, the calculation of FAR would be performed using (7.1), with N_{SA} and N_{FA} again being the total number of samples accepted and incorrectly accepted, respectively; while the calculation of FRR would be performed using (7.2), with N_{SR} and N_{FR} again being the total number of samples rejected and incorrectly rejected, respectively. This extension of *verification* mode metrics could likewise be applied to the calculation of the *identification* mode's false alarm and detection and identification rates. Moreover, we may also wish to consider the system *usage rates* when evaluating the effectiveness of a gait-based recognition system; for instance, a system with a high *failure to enroll* (FTE) rate might be considered to be less effective even if it offers desirably low error rates.

$$\text{FAR} = \frac{N_{\text{FA}}}{N_{\text{SA}}}(1 - \text{FTA}) \times 100 \qquad (7.1)$$

$$\text{FRR} = \left(\text{FTA} + \frac{N_{\text{FR}}}{N_{\text{SR}}}(1 - \text{FTA}) \right) \times 100 \qquad (7.2)$$

Building upon the error rate-based evaluation of gait biometric systems, to achieve more effective evaluations we should revisit the *Equal Error Rate* (EER), the *verification* mode rate at which the FAR is *equal* to the FRR; this metric provides perhaps the single most useful measure of biometric system performance. In Chap. 2, there was mention of how this metric could be acquired using a scalable *threshold of acceptance*. In the previous chapter, we demonstrated that the biometric system classifier component can be configured to estimate *probabilities* of samples matching identities rather than *binary acceptance* outputs. This makes it possible to set an acceptance threshold and have every returned verification probability *greater than or equal to* the acceptance threshold *accepted* by the classifier, with those *less than* the threshold *rejected*. Comparing the accepted and rejected

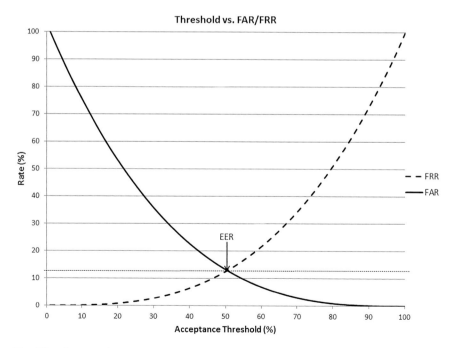

Fig. 7.2 Diagram of threshold versus EER. This diagram demonstrates how an EER can be obtained by adjusting the classifier *acceptance threshold*. In this case the EER is found when the acceptance threshold is set to about 50 %

verification requests with the expected results exposes the FAR and FRR, and, with a *variable* acceptance threshold, it is possible to tune the results in such a way that the difference between the FAR and FRR is *minimized* and the EER can be efficiently *approximated* (see Fig. 7.2). When performing an evaluation over a fixed size dataset an even better measure of effectiveness can be accomplished using the technique known as *k-fold cross-validation* [16]. This technique, which was briefly introduced at various points in the component optimization chapters of our demonstrative experiment, involves breaking the dataset into all possible training and testing subset pairs then computing the EER across each. The final evaluated result returned by this biometric system would then contain an EER that was found by adjusting a single acceptance threshold to simultaneously minimize the difference between the FAR and FRR across each of the dataset's cross-validation sample spaces, as demonstrated in Fig. 7.3.

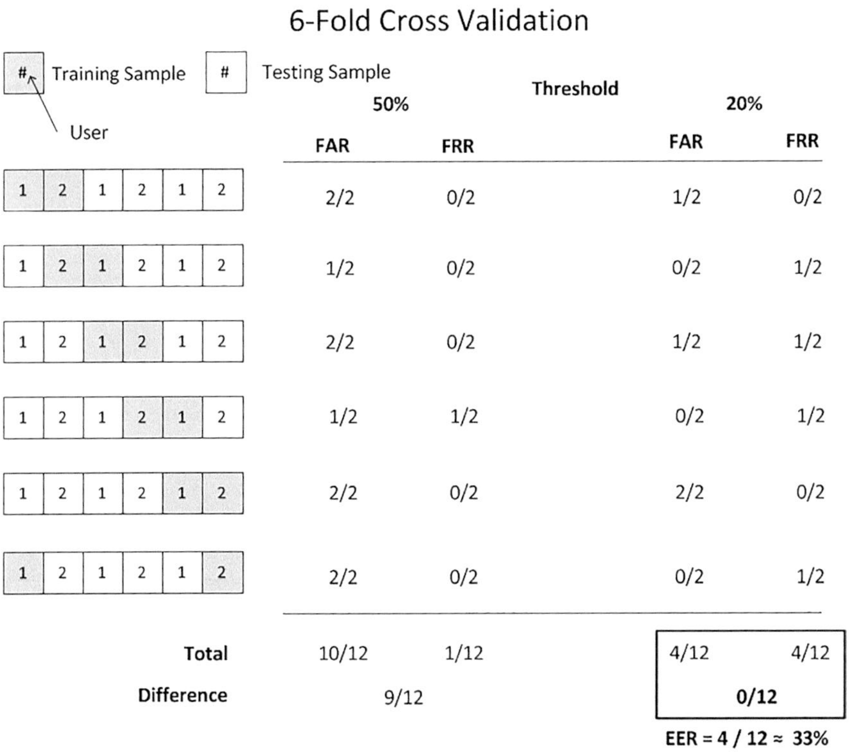

6-Fold Cross Validation

						Threshold 50%			Threshold 20%	
						FAR	FRR		FAR	FRR
1	2	1	2	1	2	2/2	0/2		1/2	0/2
1	2	1	2	1	2	1/2	0/2		0/2	1/2
1	2	1	2	1	2	2/2	0/2		1/2	1/2
1	2	1	2	1	2	1/2	1/2		0/2	1/2
1	2	1	2	1	2	2/2	0/2		2/2	0/2
1	2	1	2	1	2	2/2	0/2		0/2	1/2
Total						10/12	1/12		4/12	4/12
Difference						9/12			0/12	

EER = 4 / 12 ≈ 33%

Fig. 7.3 Cross-validation example. This figure demonstrates *k-fold cross-validation* with $k = 6$. It can be seen from this figure that evaluating a dataset using cross-validation gives us some of the benefits that could be expected when using a larger dataset. This is particularly apparent in very small datasets. In the case above, in the absence of cross-validation we could only ever estimate an EER of 25 or 50 % based on the samples chosen for training and testing. However, using cross-validation the potential probability space increases sixfold allowing us to more accurately estimate the EER value

Efficiency and *effectiveness* measure what could be categorized as the biometric system *performance*, yet, the third and final measure of the usability evaluation criteria, *user satisfaction*, measures less quantitative, possibly *emotionally* driven factors like *ease-of-use* and user *trust* in the system. These metrics are rarely studied but could have potentially significant implications on the ability to deploy a gait-based biometric system. If these are to be incorporated into the evaluation process, effective methods to acquire these metrics might involve running *user surveys* and performing *trials* to study user–system interactions.

Finally, in addition to the *data quality* and *usability* biometric evaluation criteria presented above we may also wish to consider the third evaluation criteria put forth by El-Abed et al. [3], *security*. Security for the purpose of evaluation can be difficult to quantify when compared with the other criteria; however, there are groups attempting to *quantify* security metrics for biometrics system evaluation using products such as the *Security EvaBio* platform [4]. Applying such tools to evaluate a gait biometric system may increase *confidence* in its reliability, but maintaining the security of a biometric system should be considered an ever-present challenge.

Now that we have seen some of the criteria available for evaluating gait recognition systems, the remaining sections of this chapter will demonstrate how we have chosen to perform our own evaluation of our GRF-based gait recognition experiment. This will build upon the work presented in the previous three chapters and provide the foundation for the *evaluation phase* of our demonstrative experiment, formerly discussed in Sect. 3.7 of Chap. 3.

7.2 Experimental Design

Before we can begin the evaluation phase of our demonstrative experiment we must first *design* the biometric system on which we will perform the evaluation. The objectives of this book depend on the ability to accurately collect and compare performance metrics for a variety of GRF-based gait recognition techniques. As was periodically mentioned in the preceding chapters, any biometric system used to obtain these metrics will consist of *two stages*, the *enrollment* and *challenge* phases. During the *enrollment* phase the system is provided with data to learn the biometric signatures of enrolled individuals. During the *challenge* phase, new data samples are provided to the system and recognition is performed. In Chap. 2, we saw that biometric systems can operate in one of two modes: *verification* mode or *identification* mode. For the purpose of this book, a GRF-based gait biometric system has been developed with all results acquired in *verification* mode and the primary means of evaluation being the EER *effectiveness* metric. This novel experimental biometric system has been setup to allow for multiple configurations of the *feature extraction*, *normalization*, and *classification* techniques discussed in Chaps. 4 through 6. The following subsections describe the design of this system and explain the limitations and assumptions that we have imposed upon it.

7.2.1 Recognition Techniques

The experiment performed by our GRF biometric system facilitates novel research that addresses some of the previously identified GRF-recognition research gaps. In the area of GRF normalization, two new GRF normalization techniques were designed for our research: *localized least squares regression* (LLSR) and *localized least squares regression with dynamic time warping* (LLSRDTW). These techniques are compared with the L^∞ normalization and linear time normalization (LTN) techniques used in previous GRF-recognition studies in addition to the popular L^1, L^2, and Z-score normalizers. Unlike previous studies, the research in this book also compares *normalized* to *non-normalized* GRF-recognition performance. Furthermore, it increases the GRF classification knowledge-base by extending the list of classifiers applied to GRF recognition to include the promising new *least squares probabilistic classification* (LSPC) classifier, a novel discriminative classification technique first proposed by Sugiyama in 2010 [13]. Additionally, as demonstrated in Table 7.1, the experimental research presented in this book, for the first time, compares the GRF-recognition performance of the wavelet packet feature extraction technique to the spectral and holistic techniques, the holistic feature extraction technique to the spectral, and the MLP to the LDA classifier.

As well as expanding upon previous GRF-recognition techniques, the biometric system presented here also makes it possible to compare the impact of variations in *shoe type* against each technique. And, with such information, it should be possible to identify the recognition techniques that best mitigate any of potentially negative effects that variation in shoe type might have on GRF recognition.

Table 7.1 Previously used GRF recognition techniques

Techniques	Previous GRF recognition research
Feature extractors	
Geometric	[9–12, 14]
Holistic	[1, 8, 11, 12,]
Spectral	[2, 14]
Wavelet packet	[9]
Normalizers	
L^∞	[11, 12]
LTN	[9]
L^1, L^2, Z-Score, LLSR, LLSRDTW	(Not previously studied)
Classifiers	
KNN	[2, 8–11, 14]
MLP	[14]
SVM	[9, 11, 12, 14]
LDA	[9]
LSPC	(Not previously studied)

This table provides a comparison of GRF-recognition techniques examined across previous GRF-recognition studies with those examined in this book

7.2.2 *Experimental Biometric System*

The *design* flow we decided upon for our biometric system is demonstrated in Fig. 7.4. In this figure, we can see that the process is divided into *two parts*: one using a *development dataset*, reflecting the previously shown component optimization phase of our demonstrative experiment, and the other using an *evaluation dataset*, reflecting the upcoming evaluation phase of our demonstrative experiment.

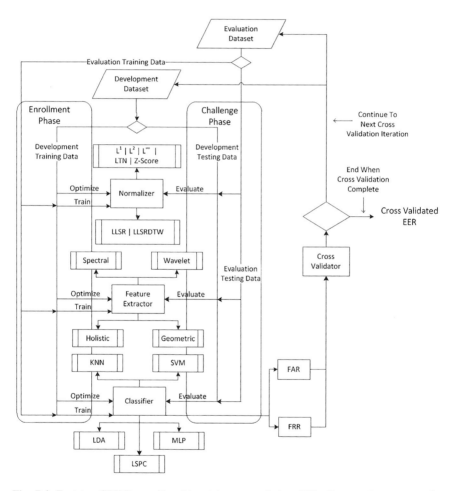

Fig. 7.4 Footstep GRF-Recognition biometric system design. This diagram demonstrates the *normalizers*, *feature extractors* and *classifiers* used by our GRF-based gait recognition system. The datasets are split into two subsets and processed in *two phases* with final classification results returned in verification mode. The *evaluation set* is only run against *development set*-optimized configurations to give results with *reduced bias* and explore alternative data formations. The process uses a *cross validator* for increased accuracy

For each of these datasets, the system was setup so that every possible configuration of the feature extractor, normalizer, and classifier could be tested. Moreover, the processing of the development dataset in Fig. 7.4 includes the additional *optimization step* performed in Chaps. 4 through 6 to discover most optimal normalizer, feature extractor, and classifier *parameters* for the purpose of GRF-based gait recognition. Using the optimized components, enrollment is accomplished by first passing the training samples through a normalizer (optionally) and feature extractor to acquire a *feature set*, and then feeding the resulting sample feature sets to a classifier so that appropriate subject boundaries can be learned. The enrollment phase is considered complete once the system is fully trained. Later, during the challenge phase, the testing subset is transformed again using the chosen normalizer and feature extractor, then all possible combinations of correct and incorrect verification requests from the testing subset are run against the trained classifier using the *k-fold cross-validation* technique discussed in Sect. 7.1.

The classifiers in this biometric system were configured to return the probability that a verification request is correct; in classifiers that did not naturally return a *posterior probability* this probability was estimated using the un-scaled output values (described in Chap. 6). The generated probability reflects the *likelihood* that a provided sample matches the given verification request, which allows us to set an acceptance threshold and, consequently, a cross-validated representation of the EER, giving us our means of evaluation. For the purpose of our *demonstrative experiment* we left the FTA out of our FAR and FRR calculations because, as we will see in Sect. 7.3, our data were collected in a laboratory setting with emphasis on capturing *complete* samples.

The implementation of the biometric system used to conduct our research was accomplished with a custom-built Microsoft Visual Studio 2010 C# solution, consisting of *four* projects. The solution included one external library and incorporated code from a number of different sources, referenced in Chaps. 4 through 6. The experimental data evaluated by this system, detailed in Sect. 7.3, consisted of a series of pre-collected GRF *stepping force* data files. A high-level architecture representative of this project is demonstrated in Fig. 7.5. The *tester* project is responsible for reading in data files, training the system, and evaluating configurations of the system using cross-validation. The *feature generator* applies the techniques used to normalize the data prior to feature extraction, as well as those used to actually extract the features. The *normalization generator* takes features extracted from the feature generator and formats (*rescales*) them appropriately for use in a classifier. To complete the classification process, the formatted features extracted from the normalization generator, are passed into a chosen classifier contained within the *classification generator* project. The final setup allows for various biometric system configurations to be compared on a single platform, reducing the potential for bias in the experimental results.

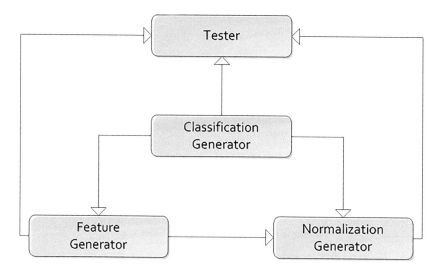

Fig. 7.5 Footstep GRF biometric system implementation. This figure demonstrates a high-level project implementation for the GRF-based gait recognition system

7.2.3 Experimental Scope

The evaluation analysis performed in this book is based on several assumptions. First, the experimental results presented in this book are applicable only under assumption that an individual's footstep GRF remains *consistent* throughout its entire recognition use period. In practical applications, this assumption may fail due to injuries, impairment, or significant differences in footwear. It has also been suggested that an individual's gait is likely to change as he or she ages [15]. Furthermore, this book defines a footstep as beginning from a *heel-plant* (or strike) and progressing in a rolling motion to a *toe push* as shown in Fig. 7.6; other variations of footstep are not considered. Moreover, footsteps, for the purpose of this book, are assumed to come from *walking* persons rather than *running* persons. When a person is walking there will always be at least one foot on the ground, however, during a running motion both feet will be off the ground for a period of time. This is important because the type of motion has a dramatic effect on the shape of the resulting GRF signature, as demonstrated in Fig. 7.7.

The research in this book also has been limited to a dataset containing only 10 different individuals. During the biometric system enrollment phase, information from each individual is used to train the system. Additionally, the data collected for this research were obtained from *cooperative* participants landing clean footsteps on a force plate. Therefore, it can be assumed that any results obtained came under nearly *ideal* conditions, and, in practice, with the potential for noncooperative and non-enrolled participants, weaker performance should be expected.

Heel-to-toe Footstep

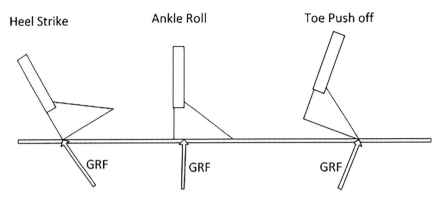

Fig. 7.6 *Heel-to-toe footstep example.* This diagram demonstrates the three primary stances during a *heel-to-toe* footstep

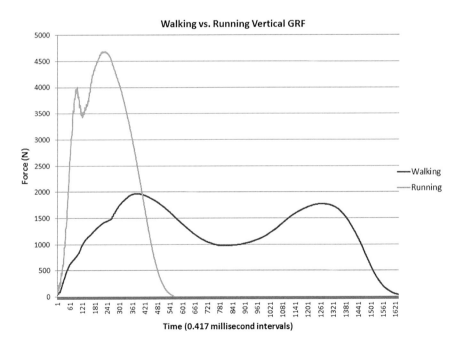

Fig. 7.7 Walking versus running vertical GRF. This figure provides a comparison of the walking to the running *vertical* GRF component in the same individual

As a final note, the GRF data collected for this research was acquired on a recent generation of *Kistler force plate*, covering the three major GRF components. While the technology may change in the future, given the Kistler's adherence to a *standard* coordinate system, we expect the findings of our research will remain relevant in any future datasets collected within the standard.

7.3 Experimental Data

The data used in this book were broken down into two smaller sets to facilitate the two parts to our demonstrative experiment. The smaller of these subsets, the *development dataset*, was used in Chaps. 4 through 6 to *optimize* our GRF-based gait recognition techniques. The larger of these subsets, the *evaluation dataset*, was established to *evaluate* the optimized biometric solutions we discovered in the previous chapters; moreover, it was made to be *mutually exclusive* of the development dataset to mitigate potential for development training *bias*. In the following chapters the performance of the various GRF-based gait recognition techniques and impact of shoe type variation will be measured against this previously unseen evaluation data. Both datasets contained footstep samples from the same 10 subjects, but varied in the shoe type used during sample collection. The development dataset consisted of 10 footstep samples per subject, taken with all subjects wearing *Asics* runners, whereas the evaluation dataset consisted of two sets with 10 footstep samples per subject, one with all subjects wearing *Orin* runners and the other with all subjects wearing *Verona* runners.

The specifications for our two datasets are demonstrated in Table 7.2. Samples were provided by the University of Calgary Faculty of Kinesiology laboratory and came from 10 different male athletes ranging in age from 21-to 30–years old. Each individual was asked to achieve a walking *speed* of approximately 1.5 m/s and make a clean step over a *Kistler force plate* [7] apparatus. The Kistler force plate used matched the rectangular plate demonstrated in Sect. 3.1 of Chap. 3 with sensors on each of its four corners. Using combinations of these sensors described in Chap. 3, the apparatus returned eight different output signals: two representing the GRF *anterior–posterior* component, two representing the GRF *medial–lateral* component, and four representing the GRF *vertical* component. The research presented in this book makes use of all eight output signals covering all three of the GRF components, which, in contrast to the previous studies shown in Table 7.3, allows the GRF dynamics to be examined in greater detail.

The graphed output of a single sample collected from the force plate is demonstrated in Fig. 7.8; while an example of the *raw* data output is shown in Table 7.4. The sample is approximately 20 seconds long and contains two distinct force spikes. Of these two force spikes, only the second spike contained useful footstep GRF information, while the first was used for an unrelated synchronization purpose. As shown in the diagram, data from the eight different output signals were collected with a 0.417 ms sampling frequency and varied between positive and negative

Table 7.2 Experimental data

Experimental data			
Apparatus	Kistler force plate		
Sampling rate	2,400 Hz		
Output signals			
GRF vertical (FZ)	GRF anterior–posterior (FY)	GRF medial–lateral (FX)	
4	2	2	
Data size	10 people		
Total samples	300		
Walking speed	1.5 m/s		
Duration	20 s		
	Samples per person	Shoe type	
Development dataset	10	Asics runner	
Evaluation dataset	10	Orin runner	
	10	Verona runner	

This table demonstrates the experiment data specifications. The eight output *signals* used in this experiment are shown according to the GRF component they belong to; there are four output signals belonging to the *vertical* component, two in the *anterior–posterior* component, and two in the *medial–lateral* component

Table 7.3 Previously used GRF components

Research group	GRF components analyzed	Data output signals
Addlesee et al. [1]	FZ	1
Orr and Abowd [10]	FX, FY, FZ	1
Cattin [2]	FZ	1
Suutala and Röning [14]	FX, FY, FZ	1
Moustakidis et al. [9]	FX, FY, FZ	3
Rodríguez et al. [11, 12]	FX, FY, FZ	1
Mostayed et al. [8]	FZ	1

This table provides a comparison of data samples components used by previous research groups

values. The output signals labeled F1X1 through F1Z4 represent the breakdown of the three GRF components, and it should be noted that the readings between the two force spikes were not entirely *smooth*, but rather, appeared to reflect small frequent *vibrations* from the floor.

To effectively analyze the various proposed GRF-recognition techniques, it was first necessary to *isolate* the desired footstep signature from the remaining extraneous data. The process of footstep *extraction* presented several challenges: first the footstep extractor needed to know when to *start* extracting the footstep, next the extractor needed to know when to *stop* extracting the footstep, and finally the extractor needed to extract only the *second* footstep rather than the first

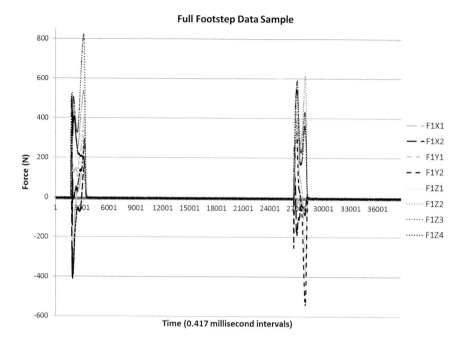

Fig. 7.8 Full footstep data sample. This figure shows the full data representation returned by the Kistler force plate, including a synchronization footstep followed by the footstep used for the research in this book

Table 7.4 Raw data sample example

Name	F1X1	F1X2	F1Y1	F1Y2	F1Z1	F1Z2	F1Z3	F1Z4
Rate	2400	2400	2400	2400	2400	2400	2400	2400
Range	10,000	10,000	10,000	10,000	10,000	10,000	10,000	10,000
0.000000	−4	−3	−3	−3	−4	0	−2	−4
0.000417	−4	−3	−4	−3	−3	−2	−2	−5
13.388750	−3	−4	−1	−2	−5	0	−3	−5
13.389167	−5	−4	14	16	−6	−2	4	0
13.389583	−8	−14	47	43	13	15	18	15

This table demonstrates an example of a segment of a raw sample output within our examined dataset. The column on the far left represents the time since the recording started (in *Seconds*), and the eight columns to the right represent the force values returned for each of the labeled sensors (in *Newtons*)

synchronization force spike. To satisfy these conditions for footstep extraction a simple process was devised.

To develop the step extractor process, it was first necessary to examine the sample data in closer detail. After examining the start and end points of several footsteps in the sample sets it was clear that all samples shared similar characteristics. It also became apparent that the force on the signals labeled F1Z1 through F1Z4 remained *positive* and *amplified* for the entire duration of a footstep. Using this finding, it was determined that the signals F1Z1 through F1Z4 would be easiest to use to establish a *threshold* for starting and stopping the extraction process.

The start and end points of a sample footstep are demonstrated in Fig. 7.9. While the start of the footstep shows a sudden well defined force spike in the Z-labeled signals, the end of the footstep is marked by a slow decline in amplitude, resulting in it being more susceptible to error from *environmental noise*. Experimental trials showed that starting the extraction of the footstep when the force in any of the Z-labeled signals exceeded a threshold of 15 N was sufficient for capturing footstep data while also generally ignoring any noise between steps. However, ending the extraction when any of the Z-labeled signals fell below the 15 N threshold was more problematic as the signals would often bounce back above the 15 N threshold shortly after, resulting in the extractor falsely believing that a new step had occurred. To account for this residual trailing noise it was determined through experimentation, that, once all Z-labeled signals fell below the 15 N threshold and all *failed* to rise above the threshold after 15 sampling intervals, the extraction process could safely be terminated.

The footstep extraction process applied to extract the samples used in both our previously demonstrated *component optimization* and upcoming *evaluation* experiments is described in the steps below. During experimentation, one sample was found to have random brief spikes of force, so an extra step (Step 3) was added to discard data when these occurred.

(1) Loop through every row in the provided data sample. For each row, if any FZ output signal is greater than 15 N then start recording.
(2) While recording, if all FZ signals have remained less than 15 N for 15 iterations, then stop recording.
(3) Discard any recording that ends up less than 100 rows long (these are likely due to error and actual steps will be around 1,600 rows long).

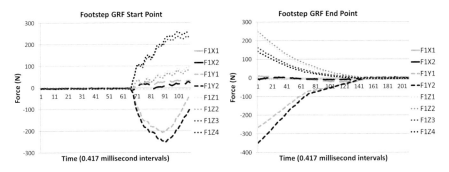

Fig. 7.9 Footstep GRF start and end points. This figure provides an example of the beginning and end of the footstep GRF

Table 7.5 Experimental footstep data parameters

	m21_1	m22_1	m22_2	m22_3	m25_1	m25_2	m25_3	m27_1	m28_1	m30_1
Asics	10	10	10	10	10	10	10	10	10	10
Verona	10	10	10	10	10	10	10	10	10	10
Orin	10	9	10	10	10	10	10	10	10	10

Training set size 5 samples per person

This table shows the available footstep data and predetermined training subset size for this experiment. The numbers under each person refer to the number of sample footsteps acquired for that person for the corresponding shoe type

(4)   Two recordings should remain at the end of the process: the first representing the synchronization force spike and the second representing the actual footstep force spike. Discard the synchronization recording and we are left with only the data from the desired footstep sample.

After applying the above process to our experimental dataset, the extracted footsteps were labeled according to their owner's *age*, with a numeric *qualifier* appended to the end of the label to distinguish different persons belonging to the same age group (i.e., m22_2). The extraction process was able to successfully extract footsteps in all but one of the 300 available data samples (Table 7.5). The unsuccessful sample, the 10th Orin sample for the m22_1 subject, was determined to contain an incomplete footstep.

To perform recognition analysis, the dataset must be broken up into *training* and *testing* subsets. And, to avoid potential *bias*, the *size* of the training set should be the same for all tests run. In this experiment it was decided that a training set of 5 samples per person would be sufficient for assessing the impact of various recognition techniques during the component optimization phase of our demonstrative experiment; however, in the next chapter we will see that potential recognition performance benefits could be achieved using larger training sets.

7.4   Summary

This chapter introduced three different criteria under which a gait-based biometric system can be evaluated: data quality, usability, and security. From there, the chapter proceeded to present the GRF-based gait biometric system designed for the evaluation phase of our demonstrative experiment with emphasis placed on the EER, a part of the usability criteria, for evaluation. It was shown that this metric can be improved using a technique known as k-fold cross-validation and noted that the biometric system classifier components, discussed in the previous chapter, had been configured specifically for the acquisition of the EER metric. The biometric system itself was designed to incorporate varying combinations of the feature extractors, normalizers, and classifiers from the component optimization phase of our demonstrative experiment. Among these components were the novel LLSR and

LLSRDTW normalizers developed in our research, and the LSPC classifier, a classifier never before used for the purpose of GRF-based gait recognition. Furthermore, in a section devoted to the experimental problem scope, this chapter clarified the assumptions and limitations made with regards to the domain for the evaluation of our demonstrative experiment.

Having presented the experimental design, the chapter continued on to describe the dataset for which our chosen biometric system was configured to work. Following best practices, the dataset was split into a development dataset, formerly used in the optimization and analysis of each biometric component, and a mutually exclusive evaluation dataset to be used in obtaining evaluation results less influenced by training bias. The two chapters that follow focus on testing our biometric system, verifying our objective assumptions with respect to the impacts of shoe type and normalization on GRF-based gait recognition and providing a more comprehensive comparison of the recognition techniques than done in any previous related work.

References

1. Addlesee, Michael D., Alan Jones, Finnbar Livesey, and Ferdinando Samaria. 1997. The ORL active floor [sensor system]. *IEEE Personal Communications* 4(5): 35–41.
2. Cattin, Philippe C. 2002. Biometric authentication system using human gait. PhD Thesis. Swiss Federal Institute of Technology, Zurich, Switzerland.
3. El-Abed, Mohamad, Christophe Charrier, and Christophe Rosenberger. 2012. Evaluation of biometric systems. *International Journal of Biometrics* 4(3): 265–290.
4. El-Abed, Mohamad, Patrick Lacharme, and Christophe Rosenberger. 2012. Security evabio: An analysis tool for the security evaluation of biometric authentication systems. In *5th IAPR/IEEE International Conference on Biometrics (ICB)*, 1–6. New Delhi.
5. ISO 13407. 1999. Human centred design process for interactive systems.
6. ISO/IEC 29794–1. 2006. Biometric quality framework standard, first ed., jtc1/sc37/working group 3.
7. Kistler force plate formulae. http://isbweb.org/software/movanal/vaughan/kistler.pdf.
8. Mostayed, Ahmed, Sikyung Kim, Mohammad Mynuddin Gani Mazumder, and Se Jin Park. 2008. Foot step based person identification using histogram similarity and wavelet decomposition. In *International Conference on Information Security and Assurance*, 307–311. Busan.
9. Moustakidis, Serafeim P., John B. Theocharis, and Giannis Giakas. 2008. Subject recognition based on ground reaction force measurements of gait signals. *IEEE Transactions on Systems, Man, and Cybernetics-Part B: Cybernetics* 38(6): 1476–1485.
10. Orr, Robert J., and Gregory D. Abowd. 2000. The smart floor: A mechanism for natural user identification and tracking. In *CHI '00 Conference on Human Factors in Computer Systems*, 275–276. The Hague.
11. Rodríguez, Rubén Vera, Nicholas W. D. Evans, Richard P. Lewis, Benoit Fauve, and John S. D. Mason. 2007. An experimental study on the feasibility of footsteps as a biometric. In *15th European Signal Processing Conference (EUSIPCO 2007)*, 748–752. Poznan.
12. Rodríguez, Rubén Vera, John S.D. Mason, and Nicholas W.D. Evans. 2008. Footstep recognition for a smart home environment. *International Journal of Smart Home* 2(2): 95–110.

13. Sugiyama, Masashi. 2010. Superfast-trainable multi-class probabilistic classifier by least-squares posterior fitting. *IEICE Transaction on Information and Systems* E93-D(10): 2690–2701.
14. Suutala, Jaakko, and Juha Röning. 2008. Methods for person identification on a pressure-sensitive floor: Experiments with multiple classifiers and reject option. *Information Fusion Journal, Special Issue on Applications of Ensemble Methods* 9(1): 21–40.
15. Winter, David A., Aftab E. Patla, James S. Frank, and Sharon E. Walt. 1990. Biomechanical walking pattern changes in the fit and healthy elderly. *Physical Therapy Journal of American Physical Therapy Association* 70(6): 340–347.
16. Zhang, Feng. 2011. Cross-validation and regression analysis in high dimensional sparse linear models. PhD Thesis. Stanford University, Standford, California.

Chapter 8
Measured Performance

In the previous chapter, the concept of separating data into an evaluation and development dataset for optimization and testing was discussed. In Chaps. 4 through 6 we saw the role a development dataset can play in optimizing biometric system components for best recognition performance. Yet, optimizing a biometric system to a development dataset may expose it to inherent *data bias* and development results may lead to the selection of biometric systems that perform poorly in practice. To be certain a particular biometric system configuration is suitable for practical use it should be put through an *evaluation* process, whereby it is tested against previously unseen data. As part of our demonstrative GRF-based gait recognition experiment, this chapter puts the best biometric configurations discovered in earlier chapters to the test by applying them to the previously *unseen* samples from an evaluation dataset.

In the sections that follow, our *evaluation dataset* results will be presented from three different perspectives. The first section presents a generalized assessment of our GRF-based gait recognition performance when applied over our *entire* evaluation dataset for our best performing classifiers from Chap. 6, while the two sections that follow examine the effectiveness of *stepping speed*-based normalization and influence of *shoe type*, respectively. In addition to demonstrating the utility of performing an evaluation over a gait recognition system, another aim of this chapter is the verification of the assertions that normalization can mitigate stepping speed differences and that shoe type variations have a negative impact on GRF-based gait recognition performance; two aspects of gait recognition that were found in Chap. 3 have received little research attention to date.

8.1 Evaluation Dataset

In Chap. 6, we derived a set of biometric system configurations to achieve optimal GRF-recognition performance when computed over our development dataset. Although these configurations demonstrated powerful recognition capabilities, the process of optimization has a tendency to *overfit* specific characteristics reflecting the dataset on which the optimization was performed. Consequently, the optimal

J.E. Mason et al., *Machine Learning Techniques for Gait Biometric Recognition*, DOI 10.1007/978-3-319-29088-1_8 configurations discovered in Chap. 6 may be far from optimal when applied in a more general setting to previously *unseen* real-world data. To better gauge the effectiveness of our biometric system configurations, in this section and the sections that follow we reapply some of our best performing experimental biometric configurations to our evaluation dataset.

Our evaluation dataset, as described in the previous chapter, was made up of 199 samples from 10 different subjects, divided according to two different *shoe types* (100 collected with subjects wearing a Verona runner and 99 collected with subjects wearing an Orin runner). Samples were split evenly across shoe types with all but one subject having 10 steps per shoe type; the *outlier* subject (m22_1) was missing a single Orin sample. For the purpose of our research, we opted to use same *k-fold cross-validation* and the EER metric to assess GRF-recognition performance over our evaluation dataset as was used to assess our development dataset performance during component optimization; however, the missing sample introduced some complexity to this approach. To account for the missing step and still use k-fold cross-validation, we would need to either *ignore* the training folds that lined up with the missing sample, pull in an *additional* sample from the testing set to replace the missing sample when needed, or allow the subject with the missing sample to be trained with *one fewer* sample than his peers. For the purpose of our research, we opted for the final of the three options and once again went with five training samples per subject in our tests. As a result of this decision, the tests that included the Orin shoe type for training could be expected to perform slightly worse than those that were purely trained with the Verona shoe type, on account of the one subject being trained on only four samples rather than five samples during some of the cross-validation folds.

While applying the biometric system configurations derived from our development system in the previous chapters, we also faced a problem regarding which of the formerly computed machine learning-based *preprocessor transformations* (i.e., PCA, WPD, LLSRDTW) from our component optimization to use in our evaluation testing. To this point, a different transformation was computed for *each* cross-validation *fold* over the development data and none of these stood out as a single best candidate to reuse in evaluation testing. Re-deriving these transformations with our evaluation dataset would introduce bias and thus defeat the point of having an evaluation set; instead, for the purpose of our evaluation testing, using the *parameters* acquired during the component optimization, we recomputed all required transformations across the full 100 samples of our development dataset and used these newly derived transformations in each of our tests as required.

The biometric system *evaluation experiment* described in this section covers *two* different experiments that we performed to provide a *general* assessment of the strength of our optimal biometric system configurations; the first covered *classification performance* with respect to preprocessors and the second examined whether performance gains could be achieved by increasing the *number of training samples*. For our first experiment, we calculated the EER for each of our best performing normalized and non-normalized classifiers from Chap. 6 over our entire evaluation dataset. This meant each of our tests involved performing classification against *both*

Table 8.1 Evaluation dataset results

	KNN	MLP	SVM	LDA	LSPC
Geometric	4.9711	5.2838	4.463	4.4909	4.2843
Normalized geometric	5.9668	4.785	4.7105	3.7316	5.399
Holistic	4.7515	3.4412	4.0201	3.19	4.9339
Normalized holistic	5.6858	4.8669	6.6368	4.1689	4.9078
Spectral	20.2084	16.9458	21.1427	15.416	22.3115
Normalized spectral	11.6936	8.5892	11.7462	5.9259	12.8159
Wavelet	3.1546	3.1081	2.9145	1.7699	2.4046
Normalized wavelet	2.8475	2.0137	4.2099	1.6434	2.4325

This table compares the EER percentages obtained for each of the best classifier-preprocessor combinations from Chap. 6

shoe types and in some cases our classifiers were trained with samples from *multiple* shoe types; as a result of this, we would expect the performance to be slightly worse than might be expected were the training and testing performed with a single shoe type (see Sect. 8.3 for more). The results obtained after performing this

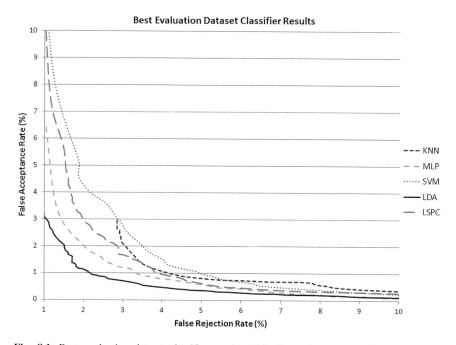

Fig. 8.1 Best evaluation dataset classifier results. This figure demonstrates the *DET curves* obtained in calculating the EER for the top performing classifier-preprocessor combinations from Table 8.1. In this figure, the KNN result appears to be cut off; this was due to a sharp near-zero *threshold*, which made it difficult to get a smooth error rate curve when approaching an FRR of zero

experiment are demonstrated in Table 8.1. These results revealed a clear decrease in performance when compared with the results obtained over the development dataset; this decrease was particularly pronounced over our *geometric* and *spectral* feature spaces, suggesting a higher degree of *preprocessor overfitting* occurred in the optimization of these spaces. With regards to classifier performance, the results were more varied; the LDA classifier performed best when compared to all others, while the SVM classifier was a bit weaker than would be expected, perhaps also owing to parameter overfitting on the development dataset. In Fig. 8.1 the differences in classifier performance can be seen more clearly with the DET curves for the best performing classifier configurations plotted against each other.

In our second experiment, we took the best *feature space-classifier* combinations from Table 8.1 and trained them using a varying number of samples per subject. To continue using our chosen cross-validation technique this experiment required that at least *two steps* be used during training, otherwise in some cross-validation folds our subject m22_1 would not be assigned any training samples, breaking the test; this training sample requirement also was bounded by our LDA classifier, which required at least two samples be present to correctly derive its *inter*-subject variance properties. Consequently, we tested the GRF-recognition performance with the number of training steps starting at 2 and increasing to 10. When plotted (Fig. 8.2) the relationship between the *number of training steps* and the GRF recognition

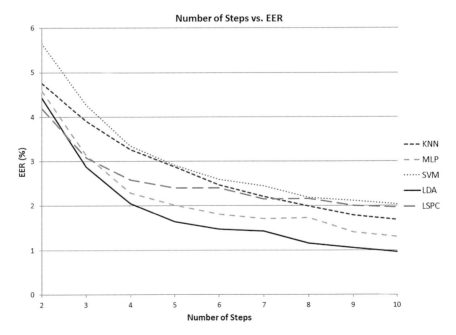

Fig. 8.2 Number of steps versus. EER. This figure demonstrates how the EER for each of our best performing classifier-preprocessor combinations changes with the number of footsteps used to train the classifier

performance became obvious, with the classifiers achieving a sharp increase in performance up to around steps 4 through 6, with slower performance increases thereafter. This, together with the fact that some of our biometric system configurations may have overfit the development dataset, could open the possibility of significantly better GRF-recognition results in larger development datasets (i.e., a dataset with enough sample variety to mitigate optimization overfitting) and/or evaluation datasets with a greater number of training samples per subject available.

8.2   Stepping Speed Normalization

One of our key research objectives outlined in the first chapter involved proving the assertion that a relationship useful for GRF-based gait recognition exists between *stepping speed* and the *shape* of the GRF force signature. In Chap. 5 we introduced several normalizers (LTN, LLSR, and LLSRDTW) designed to transform samples with varying *step durations* to the forms they would be expected to take were they all captured on a *common scale* with regards to step duration. Two of these normalizers, the novel LLSR and LLSRDTW normalizers went a step further than simply performing a common uniform transformation and attempted the acquisition of higher resolution *temporal-force* relationship models *parameterized* by the total step duration. The preliminary results calculated over our development dataset in Chap. 5 indicated that the stepping speed-force signature relationship could indeed be used to increase recognition performance; however, as discovered in the previous section, our development results were subject to an *optimization bias* and conclusions might not hold for real world data. In this section, we put these normalizers to the test against the previously unseen data of our evaluation dataset to better assess this assertion.

For the purpose of this experiment, we took the best biometric system configurations for each *step duration normalizer/feature space* pair and measured their performance across our entire evaluation dataset, once again implementing the *cross-validated* testing strategy described in the previous section. To avoid any potential optimization biases, we stuck to only computing and comparing results using a KNN classifier with a K value of 5, as this was a classifier for which all normalizers in Chap. 5 were optimized. In this experiment, we divided our results according to feature space and presented them in the form of DET curves. This representation of the results allowed us to better identify comparative performance in contrast with what could be identified from the EER alone; namely in some cases, the point at which the EER is calculated may be distorted by an anomalously *high* or *low* spike in value that would not be reflective of the performance, we would normally expect. A classifier with strong performance would be expected to have its DET curve *generally* fall under that of the others in addition to having a low EER value.

The first set of results we calculated for the purpose of this experiment was a comparison of our step duration normalizers over the geometric feature space. It must be noted that the LLSRDTW normalizer was *not* applicable to the *geometric* feature space because the heuristically obtained geometric features required no additional *alignment* (i.e., DTW) to perform regression comparisons. The DET curves that resulted from applying the remaining normalizer to the geometric feature space are demonstrated in Fig. 8.3. These results appear to strongly *contradict* our findings from Chap. 5 where the same normalizers were applied to our development dataset and normalization led to a substantial increase in recognition performance over the non-normalized data. This might suggest a higher degree of development dataset overfitting occurred in the normalized spaces over the non-normalized spaces, but it could also suggest that the step duration-based geometric features played a larger role in discriminating subjects in the evaluation dataset; such a factor could be undesirable were this to be applied to a larger dataset with more step duration-timing crossover between different users.

Our second set of results was calculated by rerunning the aforementioned normalizer performance comparisons on our *holistic* feature space, this time with the applicable LLSRDTW normalizer included. These results came out more in line with what we expected. The DET curves in Fig. 8.4 demonstrated a clear increase in GRF recognition performance when any of the step duration-normalized classifiers

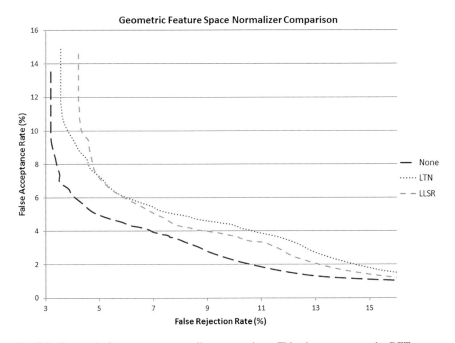

Fig. 8.3 Geometric feature space normalizer comparison. This chart compares the DET curves obtained by performing KNN classification on our evaluation dataset over the *normalized* and *non-normalized* geometric feature spaces from Chap. 5

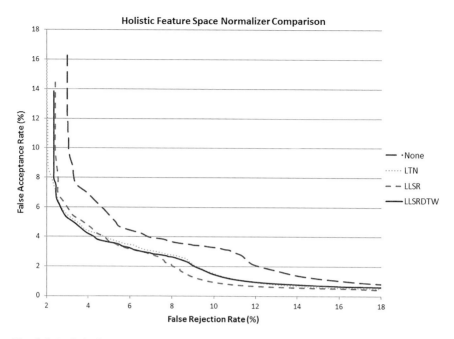

Fig. 8.4 Holistic feature space normalizer comparison. This chart compares the DET curves obtained by performing KNN classification on our evaluation dataset over the *normalized* and *non-normalized* holistic feature spaces from Chap. 5

were applied. The fact that this was in contrast to our geometric feature space results may be due to the fact that the holistic feature set contained no features directly measuring the *temporal properties* of the data. These results also stood in contrast to the normalized holistic result from the previous section. In the preceding section's *general* biometric system evaluation, normalization was performed using an L^1 amplitude normalizer, the *best* performing holistic feature space normalizer over our development dataset. The results obtained in this section would seem to imply that step duration-based normalizers can achieve better performance than amplitude scaling normalizers when applied in more complicated datasets like our evaluation set.

The third set of results we collected was taken from our *spectral* feature space. In this case the LTN normalizer was not applicable as the *time dimension* upon which such normalization is applied gets negated when the *derivative* is acquired during the transformation of raw data into our spectral feature space. The DET curves that resulted from running our cross-validated test strategy on the remaining normalizers are demonstrated in Fig. 8.5. As was the case for our general experiments in the previous section, the results obtained from our spectral feature space in this section proved to be far worse than the equivalent results on our development dataset. However, although none of our results were particularly strong, the results obtained when a step duration-based normalization technique was applied again proved to be substantially better than those when no normalization was applied.

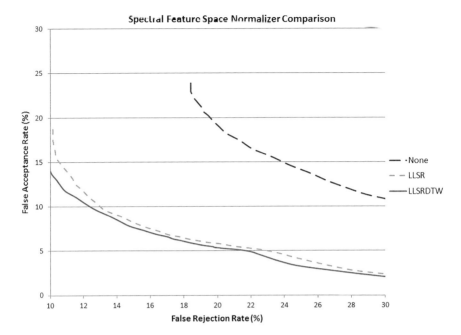

Fig. 8.5 Spectral feature space normalizer comparison. This chart compares the DET curves obtained by performing KNN classification on our evaluation dataset over the normalized and non-normalized spectral feature spaces from Chap. 5

For our final performance comparison we compared the performance of our normalizers when applied to our *wavelet* feature space. The results obtained, shown in Fig. 8.6, were considerably stronger than those obtained in other feature sets. Moreover, we once again found all of our step duration-based normalizers led to a significant increase in footstep GRF recognition performance when compared with the performance obtained when no normalizer was used.

In Table 8.2, we performed a side-by-side comparison of the EERs acquired from each of the examined feature space-normalizer combinations. Interestingly, while normalization was of no help when applied to our heuristically derived *geometric* features, all of our step duration-based normalization approaches led to an improvement in performance when applied to our *machine learning*-derived feature sets. These findings would appear to back for our assertion that normalizing for stepping speed can help improve GRF recognition performance, at least when done on a *non-heuristically* derived feature space, opening up the potential for further gains for normalization models that better fit the relationship between *force signature* and *step duration*.

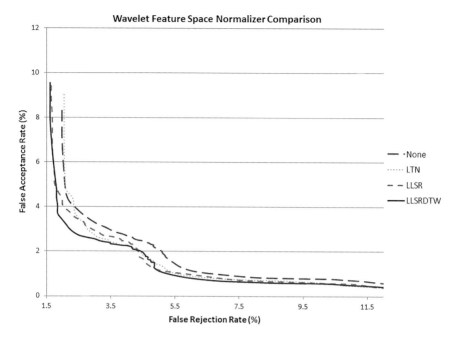

Fig. 8.6 Wavelet feature space normalizer comparison. This chart compares the DET curves obtained by performing KNN classification on our evaluation dataset over the normalized and non-normalized wavelet feature spaces from Chap. 5

Table 8.2 Stepping speed normalization results

	None	LTN	LLSR	LLSRDTW
Geometric	4.9711	5.9743	5.9668	–
Holistic	4.7515	4.3606	4.3197	4.0889
Spectral	20.2084	–	11.9114	11.6936
Wavelet	3.1546	2.8699	2.9536	2.6633

This table compares the ERR percentages obtained for each of the feature space-normalizer combinations tested in this section

8.3 Shoe Type Variation

When assessing the performance that might be expected of GRF-based footstep recognition in a real world scenario it would typically be unreasonable to assume that the people using the system never change their footwear. Instead, we might expect footwear to change based on *weather*, *formality* of the occasion, or simply due to a normal shoe-*replacement* cycle. In creating a biometric system for the purpose of footstep recognition, it is important that we understand the implications behind working in an environment that may encounter people who *enroll* with the system using one shoe and later attempt to verify using a *different* shoe. In our first chapter's objectives, we made the assertion that variation in shoe type will have a *negative* impact on recognition performance. In this section, we test that assertion and aim to gain a better understanding of the biometric system configurations that best account for shoe variations. To accomplish this we have regenerated the results matrix from Sect. 8.1 (Table 8.1), but this time using four different subsets of the evaluation dataset.

Our evaluation dataset was divided by *shoe type*, about half the samples were collected with subjects wearing a *Verona* runner and the other with subjects wearing an *Orin* runner. To provide a full assessment of the impact of shoe type across the dataset, we devised the experimental test strategy illustrated in Fig. 8.7. Under this strategy we used the *transformations* previously acquired in our

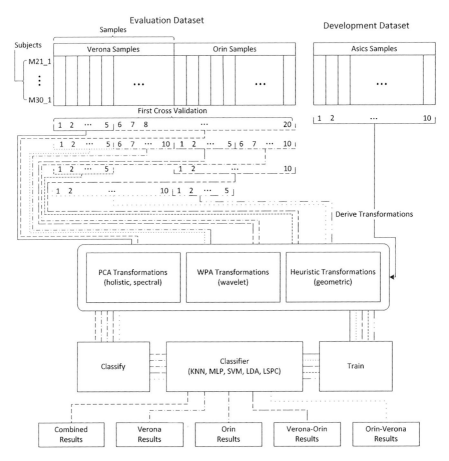

Fig. 8.7 Biometric system test strategy. This figure demonstrates the test strategy used to obtain results for our various *shoe-based* result sets. Here we provide an example of a single cross-validation with the training/testing samples identified according to their alignment to the evaluation samples

development dataset (with subjects wearing Asics runners) to perform all data preprocessing required for our evaluation testing. We already obtained the results *combining* both shoe types for training/testing (combined results) in Sect. 8.1, so for the remainder of this section we demonstrate the additional results obtained using subsets of the data across only a *single* shoe type (Verona results and Orin results) and those where the *tested* shoe type always *differed* from the *trained* shoe type (Verona–Orin results and Orin–Verona results).

The first results we collected involved classifiers *trained* with footsteps from the *same* shoe type as the type on which they were *tested*. In this experiment we reapplied the *cross-validated classifier-preprocessor* calculations from Sect. 8.1 to the 100 Verona samples (Table 8.3) and 99 Orin samples (Table 8.4) of our evaluation dataset; again, in the case of the *missing* Orin sample several cross-validations included a subject trained with only four rather than five samples. Due to the inclusion of cross-validations with a missing Orin sample we could

Table 8.3 Verona dataset results

	KNN	MLP	SVM	LDA	LSPC
Geometric	3.4222	5.1	3.7333	4.8555	4
Normalized geometric	3.8666	4.7555	4.1222	6.7333	3.0222
Holistic	3.8333	2.3777	3.1	2.8111	3.7333
Normalized holistic	4.8666	2.8555	5.7	2.9666	4.2
Spectral	10.8222	10.1666	10.0888	8.6444	11.4555
Normalized spectral	6.1888	3.0333	3.4	1.6777	5.7555
Wavelet	1.8666	1.4777	1.2	0.8111	1.0666
Normalized wavelet	1.4333	0.8	2.4555	0.2666	0.9777

This table compares the EER percentages obtained for each of the best classifier-preprocessor combinations from Chap. 6 when applied to the *Verona* shoe type samples in our evaluation dataset

Table 8.4 Orin dataset results

	KNN	MLP	SVM	LDA	LSPC
Geometric	5.0505	7.0145	4.4893	5.1402	4.2873
Normalized geometric	6.3636	6.5881	5.7126	4.3995	6.4309
Holistic	4.4893	3.1537	3.3557	2.3681	4.2985
Normalized holistic	4.2312	2.9629	5.1627	3.4118	3.1088
Spectral	16.2177	12.9854	18.6644	11.358	18.1144
Normalized spectral	10	7.7216	10.8193	4.9943	13.0639
Wavelet	3.1537	3.7485	3.0751	2.1773	2.2109
Normalized wavelet	1.8406	2.1548	3.5802	2.0875	1.1616

This table compares the EER percentages obtained for each of the best classifier-preprocessor combinations from Chap. 6 when applied to the *Orin* shoe type samples in our evaluation dataset expect a slight *negative bias* in the results obtained from that shoe type. Our expectation was also that overfitting of preprocessors and classifier parameters would result in weaker recognition performance in these sample sets when compared with our development dataset results; however, as can be seen in our Verona results table, in the case of the Verona shoe type the *normalized wavelet* results actually appeared to be nearly as strong and, in some classifiers, stronger than our development results.

Having collected our results across *single*-shoe subsets, our next experimental results were obtained by training our classifiers with the samples from *one shoe* and testing them with the samples from *the other*. In this case, the Verona–Orin result set (Table 8.5) refers to results obtained by performing *10-fold cross-validated* training with 5 samples per fold in the Verona sample set, while running each trained fold against the entire Orin sample set to calculate our EER results. The Orin–Verona result set (Table 8.6), on the other hand, refers to the *opposite*, with

Table 8.5 Verona-Orin dataset results

	KNN	MLP	SVM	LDA	LSPC
Geometric	4.9214	6.1054	5.6509	5.3928	4.6913
Normalized geometric	4.5111	4.4781	3.872	4.5735	4.8484
Holistic	4.624	3.6251	4.0235	3.973	3.9057
normalized holistic	5.4826	5.3086	6.2457	5.0392	4.9775
Spectral	23.3277	21.8013	23.5465	20.5723	25.0841
Normalized spectral	14.1526	13.3838	15.3367	11.5375	16.8855
Wavelet	3.7146	4.45	4.0123	2.6206	3.0303
Normalized wavelet	3.4511	3.0303	7.0314	2.2334	3.0808

This table compares the EER percentages obtained for each of the best classifier-preprocessor combinations from Chap. 6 when trained with our evaluation *Verona* shoe samples and tested against our evaluation *Orin* shoe samples

Table 8.6 Orin–Verona dataset results

	KNN	MLP	SVM	LDA	LSPC
Geometric	6.1833	5.2722	4.5444	4.0833	4.6722
Normalized geometric	5.9888	4.1611	5.0333	3.0444	5.1777
Holistic	4.5944	3.2333	4.1444	2.6666	5.7833
Normalized holistic	6.1277	5.8722	7.4	4.4388	5.4055
Spectral	26.6611	23.1166	28.9166	19.0444	27.75
Normalized spectral	14.5388	10.1166	15.2777	6.7166	13.3388
Wavelet	4.6722	3.6833	4.2388	1.6833	3.1944
Normalized wavelet	3.7888	2.4722	4.7166	1.5888	3.1

This table compares the EER percentages obtained for each of the best classifier-preprocessor combinations from Chap. 6 when trained with our evaluation *Orin* shoe samples and tested against our evaluation *Verona* shoe samples

Orin samples used for training and Verona samples used for testing. Again the Orin sample set may have suffered a slight *negative bias* due to a missing training sample in several cross-validations; however, in our findings the Orin-trained results often proved better than the Verona-trained results, in comparison with the previous single-shoe results where the Verona-trained results were almost universally better. Our primary expectation was that, on a whole, the results obtained by these *mixed-shoe* evaluation subsets would be worse than those obtained via the *single-shoe* evaluation subsets and this seems to have held true in these results.

The EER results from the tables presented in this section would appear to back our assertion that shoe type variation can have a *negative* impact on GRF-based gait recognition performance. To further demonstrate this, in Fig. 8.8 we plotted the DET curves for our *normalized wavelet* recognition performance over the *combined* (Sect. 8.1), *single shoe*, and *cross shoe* subsets. In this figure, we have plotted the results for each classifier type, with each appearing as a separate line identifying the shoe subset from which it was obtained. While there tended to be some overlap between these evaluation subset results due to classifier differences, on a whole we see the *solid line*-single shoe results tended toward lower EER values, the *dashed line*-cross shoe results tended toward higher EER values, and the *double*

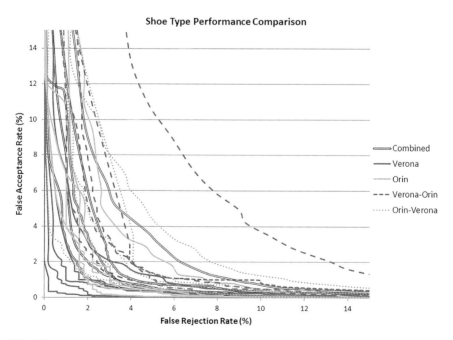

Fig. 8.8 Footwear performance comparison. This figure compares the DET curves for each *classifier-shoe* subset pair for our *normalized wavelet* preprocessor. In this case, each line represents a different set of classifier results with the line *styling* indicating the shoe subset from which it came

line-combined results tended to be somewhere in between. The position of the *combined* shoe type result curves would also support a finding in [1], which suggested training with *multiple shoes* per subject can improve recognition performance.

8.4   Summary

This chapter examined the results obtained after taking some of the top performing biometric recognition system configurations from Chaps. 4 through 6 and applying them to our evaluation dataset. While many of our results proved weaker than those obtained from our development dataset (the set on which our preprocessors and classifiers were optimized), we found that reasonably high GRF-based gait recognition rates could be obtained. This was particularly true for our wavelet feature space, which outperformed all other feature spaces. We also found that our geometric and spectral feature spaces generated results that were substantially weaker than those obtained from our development dataset, suggesting those two techniques were not very good at generalizing the recognition problem for more complicated previously unseen datasets.

With regards to the two assertions from our first chapter objectives, preliminary analysis of our results appeared to back both. We reapplied the step duration-based normalizers (LTN, LLSR, and LLSRDTW) from Chap. 5 to our evaluation data and found that, in all but our geometric feature space, each of these led to improved recognition results. Moreover, when comparing results obtained from testing against a classifier trained on samples from a common shoe type and those obtained from the same classifier tested against samples from a different shoe type, we found that the former generally performed better than the latter. In the next chapter, we wrap up the evaluation phase of our demonstrative experiment by further exploring the backing for our shoe type and normalization assertions, analyzing how our results compare with those of related research, and discussing areas that may lead to future improvements in the field.

Reference

1. Cattin, Philippe C. 2002. Biometric authentication system using human gait. Ph.D. Thesis, Swiss Federal Institute of Technology, Zurich, Switzerland.

Chapter 9
Experimental Analysis

The results of an evaluation experiment can provide insight into how a biometric system may perform in a practical setting, however, no experiment can ever capture all the complexities that may be encountered in a real-world environment. In the previous chapter, we demonstrated some key findings with respect to the performance of various machine learning techniques against GRF-based biometric data and discovered impacts of *stepping speed normalization* and *shoe type* variations on recognition performance. In this chapter, we conclude our GRF-based gait recognition experiment by *quantifying* these results and discover their significance in relation to the broader study of gait biometrics. As an important part of this process, we analyze our findings to identify potential sources of *error*, and, finally, we explore ways in which we may be able to *improve* upon the results discovered. It is hoped that this work will contribute to the present day understanding of GRF recognition and address some of the technical issues facing the deployment of such a system in a real-world setting.

9.1 Findings

The results collected in Chap. 8 appeared to support the two *assertions* of this book's objectives. A closer inspection also reveals that many of our results were consistent with findings in related work, with a few exceptions. To meet our outlined objectives, in this section, we further analyze these results and compare them with both our own experimental expectations and the findings of related research. We begin by presenting a more detailed look at how closely our results adhered to our two assertions with respect to *normalization* and *shoe type*, and then identify the areas in which our recognition results supported or contrasted the results of related research, with special attention given to our *never-before-evaluated* GRF-based biometric system configurations.

© Springer International Publishing Switzerland 2016
J.E. Mason et al., *Machine Learning Techniques for Gait Biometric Recognition*,
DOI 10.1007/978-3-319-29088-1_9

9.1.1 Shoe Type

To help verify the assertion that variation in *shoe type* can have a *negative* impact on the GRF recognition performance, the previous chapter presented the results for biometric system configurations trained and tested with the same shoe type, trained and tested with varying combinations of shoe types, and trained with one shoe type then tested against a different shoe type. By further analyzing these results, the relationship between shoe type variation and GRF recognition performance becomes clear. In Table 9.1, we compare the *averaged* and *best* classifier performance achieved across our evaluation dataset for each of our shoe type training configurations. Looking at the demonstrated comparisons, on a whole, the *same-shoe* training/testing results were stronger than the other two, while the results that included *combined* shoe types during training generally outperformed those that were obtained by always testing against a different shoe type from the one upon which the training was performed. A notable exception to this tendency was observed in the results over our *geometric* feature space, which demonstrated less performance variance across shoe types and was even shown to perform better in some cases when trained and tested across differing shoe types. This result was interesting because it was consistent with the findings of [8], in which analysis was performed over a geometric feature space and it was concluded that differences in footwear did not significantly impact the GRF recognition performance.

Table 9.1 Shoe variation findings

	Same shoe testing		Combined shoes		Cross shoe testing	
	Averaged	Best	Averaged	Best	Averaged	Best
Geometric	4.7092	3.8547	4.6986 [−0.2]	4.2843 [+11.1]	5.1517 [+9.3]	4.3873 [+13.8]
Normalized geometric	5.1994	3.7108	4.9185 [−5.4]	3.7316 [−0.5]	4.5688 [−12.1]	3.4582 [−6.8]
Holistic	3.352	2.3729	4.0673 [+21.3]	3.19 [+34.4]	4.0573 [+21]	3.1458 [+32.9]
Normalized holistic	3.9466	2.9092	5.2532 [+33.1]	4.1689 [+43.3]	5.6297 [+42.6]	4.7081 [+61.8]
Spectral	12.8517	10.0012	19.2048 [+49.4]	15.416 [+54.1]	23.982 [+86.6]	19.8083 [+98]
Normalized spectral	6.6654	3.336	10.1541 [+54.3]	5.9259 [+77.6]	13.1284 [+84.9]	9.127 [+173.5]
Wavelet	2.0787	1.4942	2.6703 [+28.4]	1.7699 [+18.4]	3.53 [+68.8]	2.1519 [+44]
Normalized holistic	1.6757	0.7141	2.6294 [+56.9]	1.6434 [+130.1]	3.4493 [+105.8]	1.9111 [+167.6]

This table presents the EER percentages obtained by comparing shoe type variations in Chap. 8 (with percent differences from the *same shoe* row results shown in the bold square brackets). The *averaged* results were collected across our five different classifiers and the *best* classifier result for *same* and *combined* shoes refers to the average of the best *individual* classifier results over both shoe types

In the related GRF recognition studies, we found three other studies describing the use of *multiple* shoe types in their dataset. These studies included [9, 10], which used multiple shoe types per person in their dataset but did not examine the *influence* of such variations on their results, and [3], where it was suggested that variation in shoe type between training and testing would have a *negative* impact on GRF performance. In [3], no direct measurement of GRF-based gait recognition performance was used to assess the impact of shoe type, however, the conclusion that was reached would also appear to be consistent with our results; in that case results were acquired using a *holistic* feature extraction technique, a technique for which we observed a substantial decline in recognition performance when training/testing shoe variation was introduced. On average, our recognition results across the full set of preprocessors and classifiers came out about 50 % worse when tested across shoe types, and, while it may be possible to generate feature spaces *less susceptible* to shoe variation, as our normalized geometric feature space turned out to be, these spaces ended up producing results that were far worse than our better performing *machine learning*-based extraction techniques. In view of this, we believe that our findings, in a general sense, do in fact verify the assertion that shoe type variation between training and testing will negatively impact GRF recognition performance, but with a noteworthy caveat that will be explored in Sect. 9.2.1.

9.1.2 Normalization

In addition to providing a backing for our shoe variation assertion, Chap. 8 also presented a set of results to help verify our second assertion regarding *stepping speed*. In that assertion we predicted a relationship between stepping speed and the GRF force signature could be obtained and successfully applied to improve GRF recognition performance. This assertion had support from the work of an MV-based gait study [2], in which the use of the LTN normalizer led to improvements in recognition performance from 8 to 20 %. In our research, we expected a similar normalization technique, this time based on *step duration* that could be used to increase performance. We tested four different normalizers including two novel ones developed for the purpose of our research (LLSR and LLSRDTW). The results obtained are shown in Table 9.2.

The improvement in the recognition performance observed for many of our step-based normalizers was in line with the improvement in the performance observed for the MV-based gait recognition in [2]. While no other previous GRF-based recognition studies examined the effect of such normalizers directly, in [7] the dataset was broken up according to three categories of stepping speed (*low*, *normal*, and *high*) and experiments were performed to compare the effect of stepping speed on GRF recognition performance. What they found was a decrease in performance of about 9–15 % when testing a *wavelet* feature space-*SVM* classifier

Table 9.2 Normalization findings

Development normalization findings

	None	LTN	LLSR	LLSRDTW
Geometric	1.3333	1.0333 [−22.5]	0.1777 [−86.6]	–
Holistic	2.5555	2.3 [−9.9]	2.5333 [−0.8]	2.3 [−9.9]
Spectral	2.0222	–	2.6 [+28.5]	1.8444 [−8.7]
Wavelet	1.2888	1.1 [−14.6]	2.3333 [+81.1]	1.4555 [+12.9]

Evaluation normalization findings

	None	LTN	LLSR	LLSRDTW
Geometric	4.9711	5.9743 [+20.1]	5.9668 [+20]	–
Holistic	4.7515	4.3606 [−8.2]	4.3197 [−9]	4.0889 [−13.9]
Spectral	20.2084	–	11.9114 [−41]	11.6936 [−42.1]
Wavelet	3.1546	2.8699 [−9]	2.9536 [−6.3]	2.6633 [−15.5]

These tables demonstrate a comparison of the EER percentages obtained after applying various step-based normalizers to our four feature spaces (with the percent differences from the non-normalized row results shown in the bold square brackets)

biometric system configuration against different stepping speeds from those on which it was trained, and a decrease of about 39–57 % when running the same tests using a *geometric* feature space. Smaller decreases in performance were observed when training samples of mixed speeds were used during training, but the results in [7] did appear to support the idea that there was room for improvement in GRF recognition if the relationship between stepping speed and the GRF signature could be *modeled* with some degree of accuracy.

Although the results demonstrated in [7] were acquired using an LTN normalizer, no direct comparison was made over non-normalized data so it is not known whether their application of normalization led to any improvements. In our own results we did observe an evaluation performance improvement across most of our feature spaces when LTN and other *step duration*-based normalizers were applied using our KNN classifier; this, however, was not the case for our geometric feature space, in which a decrease in performance was observed, likely owing to development dataspace *overfitting*. It must also be noted that, of all the normalization approaches examined, our new LLSRDTW normalizer was found to deliver the best overall evaluation performance increase, leading to an increase in GRF-based gait recognition performance of about 14–15 % in our better performing holistic and wavelet feature spaces. Unlike the LTN normalizer, which performed a simple linear *scaling* operation on the data, the LLSRDTW normalizer was designed to dynamically *model* the relationship between GRF force signature and stepping speed. Therefore, because it led to improved recognition performance over the LTN normalizer, we have come to the conclusion that a modeled relationship between step duration and the GRF signature can in fact be utilized to achieve better recognition results, satisfying the second of our two assertions.

9.1.3 Biometric System

As a key objective in presented in this book's introduction, we aimed to expand upon the work done in previous GRF recognition studies with respect to *feature extraction* and *classification*. Back in Sect. 7.2.1 of Chap. 7, we presented a breakdown of our studied feature extraction, normalization, and classification techniques according to their use in previous GRF recognition studies. While many studies examined one or more of our chosen biometric system configurations, none of the previous research compared recognition performance over the *entire* set. In the previous section we demonstrated our findings with regard to the *normalization* component of our biometric system. While normalization was an area of focus in this book due to its relative lack of exposure in previous research, we also uncovered some interesting findings when analyzing our various feature extraction and classification techniques.

In Table 9.3 a comparison of our various examined feature extraction techniques is presented for results that have been averaged across all tested classifiers. One of the most obvious disparities in these results was the substantial decline in recognition rates when comparing the development and evaluation *spectral* feature space results. Interestingly, this decrease in performance appears to be strongly influenced by the evaluation use of different training and testing *shoes*, because the Chap. 8 results for the *single shoe type* training and testing subsets proved considerably better. It stands to reason that some characteristics specific to shoe type may have *negatively* influenced the selection of *principal components* during the derivation of the spectral transformation on our development dataset. Aside from the spectral results, we observed a general decrease in recognition performance between the *development* and *evaluation* dataset of approximately 100–200 %. The decrease in

Table 9.3 Feature space findings

	Average development classifier results	Averaged evaluation classifier results
Geometric	2.0911	4.6986 **[+124.6]**
Normalized geometric	0.9377	4.9185 **[+424.5]**
Holistic	1.6355	4.0673 **[+148.6]**
Normalized holistic	1.6266	5.2532 **[+222.9]**
Spectral	1.5244	19.2048 **[+1159.8]**
Normalized spectral	1.2577	10.1541 **[+707.3]**
Wavelet	1.4221	2.6703 **[+87.7]**
Normalized wavelet	1.131	2.6294 **[+132.4]**

This table compares the features space-influenced EER percentage results averaged across our five different classifiers. The values in the bold square brackets represent the same-row percent difference between the two datasets recognition performance between the evaluation and development datasets was far greater than those observed in [9, 10], which also broke data into separate development and evaluation datasets; however, our datasets contained much more information per sample and overall our performance results were far stronger than the ones acquired in those studies.

Due to the differences in GRF sample characteristics between our data and that of previous studies, we were not able to *directly* compare our results with related research. Instead, to gauge our feature space performance in the context of related work, we decided to assess our results via examining the relative performance differences found when results were acquired for two or more feature spaces in the related studies. With regard to the *holistic* feature space, in our experiment we observed a 10 % improvement in GRF recognition when we compared our holistic feature space with our geometric space (using an SVM classifier). The same configuration led to a 21 % improvement in recognition results when examined in [10]. In [13] more combinations of feature space-classifier pairs were compared with demonstrated improvements of 35, 31, and 27 % when *geometric* feature spaces were used over *spectral* feature spaces for the KNN, MLP, and SVM classifiers, respectively; in our evaluation results we discovered improvements of 75, 68, and 78 % for the same respective feature space-classifier pairs. Moreover, in [7] the use of a *wavelet* feature space led to an improvement in recognition results of over 25 % when compared against a geometric feature space (via an SVM classifier); this result turned out to be similar to our own, in which case the same configuration led to a 34 % improvement in recognition performance. So, while the results in [10, 13] might suggest the holistic or geometric feature spaces were best suited for GRF recognition performance, our comparative performance over the wavelet feature space was noticeably stronger than either.

In contrast with our feature extraction findings, the results acquired from our classifiers demonstrated less variability between one another. These results, shown in Table 9.4, compare our two datasets *averaged* across the *geometric*, *holistic*, and *wavelet* feature spaces; we deliberately left the *spectral* feature space out from this analysis as it was found to produce *outlier* results. One interesting finding here was

Table 9.4 Classifier findings

Classifier results (ignoring *spectral* feature space)		
	Average development feature space results	Averaged evaluation feature space results
KNN	1.3462	4.5628 **[+238.9]**
MLP	1.7499	3.9164 **[+123.8]**
SVM	1.4407	4.4924 **[+211.8]**
LDA	1.5277	3.1657 **[+107.2]**
LSPC	1.3055	4.06 **[+210.9]**

This table compares the classifier-influenced EER percentage results averaged across our geometric, holistic, and wavelet feature spaces. The values in the bold square brackets represent the same-row percent difference between the two datasets. Note that we ignore the outlier spectral feature space in our averaging the significant difference in classifier rankings between the evaluation and development dataset. For instance, the KNN classifier, which was second best over the development data, turned out to be the worst performing classifier over the evaluation data. Also, the LDA classifier, which was second worst in the development data, turned out to be the best by a wide margin of about 19 % in the evaluation data. When compared with the findings in previous *multi-classifier* GRF recognition studies [7, 9, 13] our averaged evaluation KNN results were consistent, with KNN performing worse than SVM, MLP, or LDA in those studies. However, our findings differed from [7, 13] with respect to SVM classifier performance. In [13] the SVM classifier performed about the same as the MLP classifier, while in our findings the SVM was noticeably worse overall (though substantially better over the *geometric* space). Moreover, in [7], the SVM classifier produced similar performance when compared with the LDA classifier, whereas in our case the SVM classifier performed far worse than the LDA classifier (about 61 % worse in the equivalent *wavelet* feature space). For the purpose of our research we also examined a classifier not used in any of the previous GRF recognition studies, the LSPC classifier. The evaluation performance acquired from this classifier ended up being around the average of the other four classifiers.

In addition to the classifiers themselves, we also examined the performance effects resulting from changes in the *number of training samples*. By doubling the number of samples in training from 3 to 6 we saw a 37 % increase in *wavelet* space recognition performance, while doubling the number of samples in training from 5 to 10 led to a 32 % increase in performance. The improvements in performance from adding training samples became less significant as more samples were added, a finding that was consistent with those of [9], in which performance was found to increase with samples added up to about 40 samples and leveled off thereafter; this would also imply that our results could be improved further by increasing the number of training samples.

9.2   Considerations and Implications

The findings presented in the previous section gave us some insight into the behavior we could expect from our biometric system in an experimental environment. However, if we were to take our findings and apply them to a real-world environment, there are a number of *caveats* that must be considered. Moreover, the characteristics of the results backing our findings and the methods by which they were acquired have important implications for how we would go about selecting an appropriate biometric system configuration for deployment. In the subsections that follow we examine these considerations and implications as they relate to three different aspects of our biometric system.

9.2.1 Data

The practicality of our findings with regard to a real-world system is subject to constraints imposed upon it by our chosen dataset. Our dataset differed from any dataset used in previous GRF recognition studies, and thus our results were not *directly* comparable with those of related research. Several other GRF studies with larger datasets used data collected in a more realistic environment with more subjects, that may *not* have been as cooperative as our examined subjects. Under that assumption, we expect that our biometric system performance would be comparatively weaker than the findings might suggest when matched against the results of a study like [10]. Furthermore, our data and analysis have several *constraints* that may not be realistic in a practical setting; namely, we only examined GRF samples that included a full *heel-to-toe* step, we only examined *walking* samples, we did not include *imposter* attempts from people not seen during training, and, in our multi-shoe analysis we only examined variations of *running shoes* (ignoring other potential footwear like boots or sandals). Moreover, in the research presented in this book we did not consider the actual foot (*left* or *right*) used in obtaining the samples and further investigation would be needed to assess the effectiveness of our techniques when subject to varying feet used in training and testing. All of these concerns could potentially result in reduced performance in a practical setting; however, when accounting for different aspects of our data with respect to a real-world setting there are other ways in which our application would have a tendency to produce better results than those discovered in our findings.

One area in which our chosen data may have led to weaker results than those that could be expected in a real-world scenario relates to the *uniformity* in the selected subjects. All of our samples were collected from *young athletic males*, yet in a practical setting we would expect a larger demographic of subjects with far more *inter*-subject variability, making it easier, on average, to distinguish different subjects. Additionally, variation in footwear *between* subjects, as opposed to the *within*-subject variations examined in this book, may be more likely in a real-world application. If *inter*-subject variation in footwear turned out to be more common than *intra*-subject variations then we might expect such variation to improve upon our recognition performance with the choice of footwear helping to distinguish subjects. Finally, there is also reason to believe that the recognition improvement observed in our research with respect to previous GRF studies may have come, at least in part, as a result of the more information rich samples captured; in our work we examined 8 distinct GRF signals with a dimensionality of approximately 12,800 points per sample, a larger sample space than any of the previous GRF recognition studies.

9.2.2 Preprocessing

After analyzing our findings with respect to our data preprocessors (*normalizers* and *feature extractors*), we were able to identify several factors that may influence *results* and *usability* were we to apply our biometric system to a practical setting. One area that we chose not to analyze during the evaluation of our demonstrative experiment was the relative *computational efficiency* for each of our preprocessing techniques; yet, if the system were to be deployed in a practical setting we may find that the performance boost gained by performing LLSRDTW normalization could, for instance, be offset by the increase computation time due to its inclusion. Still, there are other ways in which a more practical setting may introduce recognition performance improvements to our findings, particularly with respect to the greater number of samples available. Many of our preprocessors were built upon *statistical analysis*-based techniques (i.e., PCA, Fuzzy c-means clustering, and regression), which we expect would produce more accurate models were more data available for their computation. In that respect, our results may also have potentially been weaker than might be expected in relation to those GRF recognition studies with more available training samples like [9, 10] or [7].

Looking further into our preprocessor findings, we were able to arrive at several additional considerations and conclusions. First and foremost, we proved that *normalizing* for *stepping speed* could in fact be done to boost recognition performance and we suspect an even greater improvement could be accomplished with the development of stronger stepping speed-GRF signature models. Looking back at our findings, we can also see that the choice of *feature extraction* technique had more of an impact on performance than the choice of classifier or normalizer, suggesting further research into feature extraction techniques might lead to greater improvements in GRF recognition than research focused on either of the other biometric system components. It should also be noted that many of our preprocessing techniques varied slightly when compared directly with the same techniques in related studies, making direct comparisons with previous research difficult; for instance, in our *wavelet* feature extractor we performed wavelet decomposition using the Legendre 04 wavelet function, which we found produced better performance than the Coiflet 06 wavelet function used in [7]. One final interesting implication related to the domain upon which the feature extraction was performed. When acquiring our *spectral* feature space, extraction was done completely within the *frequency domain* and we found that results suffered when analysis was extended to include multiple shoe types. By comparison, the *holistic* and *wavelet* feature spaces, computed over a *time* and *time-frequency* domain, respectively, resulted in relatively small decreases in performance when extended to multiple shoe types, suggesting a possible weakness in the use of frequency domain analysis when subject to environment variations during GRF acquisition.

9.2.3 Classification

The factors affecting our *classification* findings were similar to those affecting our preprocessor findings. In a practical setting *computational efficiency* would likely become more important and classifiers such as the LDA classifier, which involved large matrix transformations, might not overcome the efficiency-performance trade-off. Yet, once again, when analyzed over a *larger* more realistic dataset we may be able to achieve improved recognition performance; in this case, by performing classifier *parameter optimizations* over a larger dataset we expect we would be able to derive parameters that better generalize subject classification boundaries. Taking this into account, it again seems that our results could have been weaker, with respect to those of previous studies, than might be expected in a practical setting. Moreover, some of our classifiers refer to algorithm *variants* that differ from those studied in related work (i.e., our use of weighted KNN vs. non-weighted KNN in other studies); however, in this case it is hard to tell whether or not these variations improved or decreased our results relative to those of the previous GRF recognition studies.

In addition to analyzing classifier performance, this book categorized classifiers as being either *generative* or *discriminative*, and either *eager learning*-based or *instance*-based; we examined classifiers belonging to each of these categories. Further analysis reveals that the category a classifier belongs to may have important implications with respect to how it might perform in a practical setting. Instance-based classifiers, like the KNN classifier, retain all training data in memory and often need to access the entire set of training samples every time a classification is required; consequently, such classifiers may suffer in terms of both efficiency and memory usage as datasets grow. The use of a generative classifier, like the LDA classifier, may also adversely affect performance. In this category of classifier there is an expectation that some knowledge of class (subject) distribution in the dataset is known a priori, while in a practical setting it may be unfeasible to get a good estimate of the class distribution. Consequently, looking at the classification techniques strictly at a high level, there is reason to believe that eager learning-based discriminative classifiers may be best suited for application to the GRF-based gait recognition problem in a practical setting.

9.3 Potential Improvements

In this book, we presented a comprehensive experimental analysis of GRF-based gait recognition using *four* different feature extractors, *seven* different normalizers, and *five* different classifiers. Our findings, though subject to the caveats in the previous section, demonstrated that with an appropriate choice of feature extractor, normalizer and classifier strong recognition performance can be achieved. That being said, we believe further improvements could be gained by both expanding the *optimization* of the recognition techniques studied in this book and testing promising *alternative* recognition techniques. In the following subsections, we examine such potential improvements as they pertain to each of our biometric system components.

9.3.1 Feature Extraction

In Sect. 9.2.2 we identified the choice of feature extractor as perhaps the most important factor in achieving strong GRF recognition results. For the purpose of this book all feature extractors were optimized using the KNN classifier and, as a result of our optimization, we may have formed a *bias* toward expectations of that classifier. It then stands to reason that we may be able to improve upon our recognition performance by simply re-running the feature space optimization for each of our different classifiers to obtain results unaffected from the KNN bias. Alternatively, shifting our focus toward other feature extraction techniques, there are several other techniques that have the potential to offer improved recognition results including: *Partial Least Squares* (PLS) [12], *Kernel Principal Component Analysis* (KPCA) [11], and *Generalized Principal Component Analysis* (GPCA) [3]. Each of these techniques have roots in PCA, though two of them (PLS and GPCA) take a supervised extraction approach, applying a *weighting* based on class membership to derive their dimensionality reducing transformations, while the KPCA technique is simply the application of PCA over the *kernel space*, just as was demonstrated for our LDA classification technique in going from ULDA to KUDA.

9.3.2 Normalization

The options available to improve upon recognition performance via normalization improvements were more limited when compared with those available to feature extraction. Our work covered most of the normalizers recommended in statistics literature, and of all our examined normalizers only the LLSRDTW involved any parameter optimization, with the *Sakoe-Chiba Band* as a single optimization parameter. In our work this optimization was done using the KNN, again creating a *bias* for that particular classifier, therefore it is possible that redoing this optimization over other classifiers could potentially lead to an improvement in GRF recognition performance. Alternatively, though not quite normalization by definition, we may be able to achieve an improvement in our GRF-based gait recognition results by dropping the regression aspect of LLSRDTW and using the technique's sample *alignment* procedure as a standalone preprocessor; in this case we would be performing feature extraction directly on the *center star* aligned spaces rather than via derived warping functions. Another possibility might include using DTW and the center star algorithm to generate separate alignments for each subject to capture inherent subject-specific characteristics that might otherwise be missed in a more general modeling; again this would involve passing the aligned spaces directly into the feature extractor.

9.3.3   Classification

Classification offers far more choices for potentially improving upon recognition results than were available for the two preceding biometric system components. There are numerous classification techniques available for all sorts of specialized purposes; moreover, classifiers tend to have massive *optimization spaces* meaning whatever parameters are discovered in a standard parameter optimization are unlikely to be in their most optimal state. Classifier optimization in nearly all our classifiers was accomplished using brute force searches of arbitrarily chosen numeric intervals. Consequently, we believe improved performance could be achieved over most of the classifiers examined in this book via the incorporation of smarter parameter space optimization techniques. Another approach to improve upon current recognition performance might be to include a training sample *rejection* strategy, as was proposed in [13], or perhaps a *fusing* of different feature spaces for classification (also proposed in that same study). Alternatively, there are several other classification techniques that could potentially lead to improved recognition performance: the *Hidden Markov Model* (HMM) classifier [5], the *Regularized Linear Discriminant Analysis* (RLDA) classifier [4], and the *Max-Margin Markov Network* (M^3-net) classifier [6]. The HMM is a *generative* classification technique which was previously used for GRF recognition in [1] and is often credited for its strength in recognizing *temporal patterns*. The RLDA classifier is an LDA variant that might be expected to achieve similar results when compared to our LDA classifier, but is said to be less prone to the *overfitting* problems that impacted our tested LDA variant. Finally, the M^3-net classifier is a leading edge classifier, which incorporates concepts from both the HMM and SVM classifiers to derive classification models; it is unlikely to have ever been tested for the purpose of GRF-based gait recognition, but, if it truly benefits from the properties of both SVM and HMM, it could potentially lead to a substantial improvement in recognition results.

9.4   Summary

In this chapter we performed a thorough analysis of the results obtained in the previous chapter. We began the chapter with a detailed look at the recognition performance achieved in Chap. 8 and compared our findings with those of related GRF recognition studies. We continued on to discuss the feasibility of a deployment of our biometric system to a practical setting, identifying issues that were not within the scope of our demonstrative experiment yet would need to be addressed before attempting such a deployment. Finally, the chapter concluded with a look into some promising alternative recognition techniques together with ways in which we may be able to optimize our current recognition techniques for better performance going forward.

Key findings presented in this chapter supported our objective assertions. We discovered an approximately 50 % decrease in recognition performance as a result of footwear variations between biometric system training and testing. Moreover, a 14–15 % increase in recognition performance was observed with the use of LLSRDTW to normalize for stepping speed over our best performing holistic and wavelet feature spaces. Additionally, when comparing different feature spaces, our wavelet space produced the best results with an EER of about 2.6% averaged across all classifiers. And, when comparing different classifiers, we found that the LDA classifier achieved the best overall performance, about 19% better than the next best classifier. Yet, to deploy this biometric system to a practical setting there are issues that must be addressed, and, as discussed throughout Sect. 9.2, we believe some of our recognition techniques may be undesirable options when computational efficiency is considered. Having concluded our demonstrative experiment in this chapter, the next chapter will delve into some of the applications that could potentially make use of such a gait-based biometric system to solve real-world problems.

References

1. Addlesee, Michael D., Alan Jones, Finnbar Livesey, and Ferdinando Samaria. 1997. The ORL active floor [sensor system]. *IEEE Personal Communications* 4(5): 35–41.
2. Boulgouris, Nikolaos V., Konstantinos N. Plataniotis, and Dimitrios Hatzinakos. 2006. Gait recognition using linear time normalization. *Pattern Recognition* 39(5): 969–979.
3. Cattin, Philippe C. 2002. Biometric authentication system using human gait. Ph.D Thesis. Swiss Federal Institute of Technology, Zurich, Switzerland.
4. Friedman, Jerome H., 1989. Regularized discriminant analysis. *Journal of the American Statistical Association* 84(405): 165–175.
5. Ghahramani, Zoubin. 2001. An introduction to hidden markov models and bayesian networks. *International Journal of Pattern Recognition and Artificial Intelligence* 15(1): 9–42.
6. Julien, Simon-Lacoste. 2003. "Combining SVM with graphical models for supervised classification: an introduction to Max-Margin Markov Networks. University of California Berkeley, Berkeley, CA, USA, Project Report 2003.
7. Moustakidis, Serafeim P., John B. Theocharis, and Giannis Giakas. 2008. Subject recognition based on ground reaction force measurements of gait signals. *IEEE Transactions on Systems, Man, and Cybernetics-Part B: Cybernetics* 38(6): 1476–1485.
8. Orr, Robert J., and Abowd, Gregory D. 2000. The smart floor: A mechanism for natural user identification and tracking. In *CHI '00 Conference on Human Factors in Computer Systems*, 275–276. The Hague.
9. Rodríguez, Rubén Vera, Nicholas W. D. Evans, Richard P. Lewis, Benoit Fauve, and John S. D. Mason. 2007. An experimental study on the feasibility of footsteps as a biometric. In *15th European Signal Processing Conference (EUSIPCO 2007)*, 748–752. Poznan.
10. Rodríguez, Rubén Vera, John S. D. Mason, and Nicholas W. D. Evans. 2008. Footstep recognition for a smart home environment. *International Journal of Smart Home* 2(2): 95–110.
11. Schölkopf, Bernhard, Alexander Smola, and Klaus-Robert Müller. 1997. Kernel principal component analysis. In *Artificial Neural Networks—ICANN'97* 583–588. Berlin: Springer.

12. Schwartz, William Robson, Aniruddha Kembhavi, David Harwood, and Larry S. Davis. 2009. Human detection using partial least squares analysis. In *IEEE 12th Internation Conference on Computer Vision*, 24–31. Kyoto.
13. Suutala, Jaakko, and Juha Röning. 2008. Methods for person identification on a pressure-sensitive floor: Experiments with multiple classifiers and reject option. *Information Fusion Journal, Special Issue on Applications of Ensemble Methods 9* 9(1): 21–40.

Chapter 10
Applications of Gait Biometrics

The gait biometric, having only relatively recently become technically feasible as a means to provide security, has to-this-date seen only limited applications in industry. Nevertheless, gait biometric recognition continues to be a growing area of interest due to its unobtrusive nature and an increasing number of technologies available for capturing information about the human gait. In this chapter, we explore a variety of *applications* for which gait biometrics may be deployed. We also discuss the *modes of operation* under which these applications can be run, and we examine the current usage of the biometric, both from *commercial* and *research* perspectives. We conclude by demonstrating the ways in which gait biometrics may *complement* some existing systems and technologies. This chapter provides a context under which the research findings presented in this book may be of benefit in the ongoing pursuit of novel applications that utilize the gait biometric.

10.1 Application Areas

As a biometric technology, gait recognition can be used for *security* and *forensics* applications. Beyond these areas of application, *medical* applications of gait recognition are also being envisioned, for instance, gait rehabilitation for patients suffering from paralysis or sport injuries. In this chapter, our focus is on the *security* applications of gait recognition technology.

Gait biometrics can be collected *unobtrusively* without forcing the subject to perform specific actions. In particular, unlike other biometric technologies, video-based gait systems do not require any direct contact with the subject. The samples can be collected and processed remotely without active cooperation of the subject. Wearable gait technologies, in contrast, require some contact with the subject. This constraint has an impact on the type of applications covered by such technologies. While gait is still emerging as a biometric technology, various applications are currently being envisioned, including the following:

- Automated surveillance and monitoring for high-security facilities
- Access control for restricted areas and sites, devices, and services

© Springer International Publishing Switzerland 2016

J.E. Mason et al., *Machine Learning Techniques for Gait Biometric Recognition*,
DOI 10.1007/978-3-319-29088-1_10

- Mobile device security
- Home automation

Gait recognition can be used to secure and control the *access* to protected facilities and devices, and for automated airport security checks. For instance, in busy airports and border crossings, it can be used for early identification and categorization of travelers (for example, using databases such as the no-fly list). Gait recognition can also be used to *reinforce* ATM security, by combining the traditional PIN with additional factors based on the gait.

GRF-based gait recognition can be used for *access control* to and within buildings and homes. For instance, the sensors can be integrated in rugs or areas at the entrances and other locations in buildings and homes. Based on the identity of the subject, access to the home or to specific rooms will be granted or denied. The same approach can also be used to control the access to specific devices or features in the home based on the identity of the individual. Subjects accessing a high-security facility might also be monitored by *continuously* authenticating them using gait recognition.

10.2   Modes of Operation

Biometric recognition systems rely on generic *modes of operation*, which sit at the core of any field applications of these technologies (Fig. 10. 1).

The following generic mode operations can be associated with gait recognition systems:

- Authentication
  - Static
  - Reauthentication
  - Continuous authentication
- Identification

Gait recognition can be used to *reauthenticate* an individual in motion within a protected area. Part of this authentication would rely on some identity claimed by the subject. For instance, this can be provided by a magnetic identification card borne by the individual. Based on the *claimed* identity, gait biometric authentication can take place. Authentication is a one-to-one verification process, where the sample gait captured is compared against the biometric *template* corresponding to the claimed identity.

Continuous authentication provides another means to identify an individual. It consists of repeating the authentication process successively at different time intervals or based on a predefined amount of the received gait samples. Continuous authentication also relies on the identity initially claimed by the subject. Subsequent verifications are based on this initial identity. A failed authentication can lead to the generation of an *alarm* and, in response to this, an appropriate action can be taken.

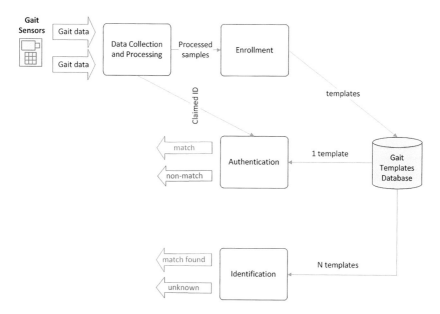

Fig. 10.1 Biometric modes of operation. This figure presents different modes of operations used for gait biometric applications

Since there is a possibility for a false alarm, the response could be gradual, and could consist of further verification such as face recognition.

While *video*-based recognition may be suitable for continuous authentication, it may not be adequate for access control to a particular area. In this case, the subject must be authenticated *statically* at the entry point of a protected area. Ground-reaction-based gait recognition is more appropriate in this scenario since sensors can be embedded in the entrance floor.

Identification applies verification in a one-to-many setting. In this case, the identity of the subject is unknown. The samples captured from the individual are compared one-by-one against the biometric profiles from known individuals previously collected and stored in a database (in forensics these could be past offenders). The comparison will yield a *match* or *non-match*. In some cases, several close matches may be returned.

10.3   Gait Biometric Methods in Commercial Use and Research and Development

To date, gait biometrics technologies have received only a limited exposure in a *commercial* setting, yet a number of studies have revealed the potential for the use of gait biometrics in industry. In an early study for the Atlanta Business Chronicle,

Barry [1] identified the potential *security* applications for *video* and *radar*-based gait recognition approaches while also noting that these approaches would still be a long way from being commercially applicable. More recently, such recognition has come closer to a commercial feasibility. As example, Panchumarthy et al. [10] produced a *case study* that performed an actual *cost-time evaluation* of video-based gait recognition algorithms using Amazon Web Services for *cloud* computations. In the same vein, perhaps the greatest driver of the commercial interest in gait bio-metrics was a program sponsored by DARPA that led to the establishment of the *HumanID Gait Challenge Problem* [13], an attempt at providing a foundation upon which video-based gait recognition can be assessed. The associated *challenge* included: the establishment of a large dataset of video walking sequences available for use by the research community, a baseline algorithm that provides a perfor-mance benchmark, and a set of 12 testable experiments with varying degrees of difficulty. This project provided an important contribution to the research com-munity as special care was given to create a *challenge* reflecting the conditions (*covariate factors*) that may be expected when dealing with real-world applications. Unfortunately, so far the commercial applications of video-based gait recognition appear to have been languishing in their development due to legal and technical implementation barriers.

Interestingly, while the aforementioned 'at a distance' *machine vision* (MV)-based gait recognition approaches have received by far the most research attention, the first actual commercial incorporation of gait biometrics has come from the *wearable* and *floor* sensor industries. Compared to the video-based approaches, the sensor-based approaches have lower barriers for commercial entry because pro-grammable wearable devices are already widely used and behavioral monitoring such as gait detection appears to be viewed as *acceptable* by the consenting owners of such devices. Consequently, several companies are currently in the process of launching gait biometric-based products. Examples of such companies include Plantiga [11], which is developing a shoe-based gait tracking system, Zikto [15], which is developing a fitness band with built-in gait biometric tracking, and ViTRAK Systems Inc. [14], which has incorporated the gait biometrics as an application of its Stepscan floor sensor system, to name a few. Another sensor-based development with commercial implications is the a mobile security solution jointly being developed by Boeing and HRL Laboratories and funded by the United States Department of Homeland Security [6]. This project is meant to utilize the aspects of gait biometrics to provide a continuous monitoring of a user's behavioral patterns while he/she is in contact with a specialized phone. It is envisaged that sensor-based approaches will gain more popularity in the near future because they provide more possibilities in terms of integration with other forms of biometrics for human identification purposes. This aspect of gait biometrics is discussed in the next section.

10.4 Integration of Gait Biometrics with Other Biometrics or Technologies

Unlike most other biometrics, the gait biometric is typically implemented in an *unobtrusive* manner such that a subject does not need to alter his/her behavior for a sample capture. Consequently, only a limited number of additional biometrics can be integrated with the gait biometrics without introducing obtrusiveness [2]. Where proper equipment is available, one method for accomplishing a *multimodal* gait recognition might involve the integration of two different forms of gait biometrics. For instance, in Cattin [3] an integration of the floor sensor (FS) with the machine vision (MV) approach is proposed, yielding an improved gait biometric approach compared with the standalone approaches. Another popular biometric integration approach is the combination of *gait* and *facial*-recognition [5, 7], which often requires no additional sensors than would otherwise be needed to capture the gait video sequences. Various techniques for performing this type of *integrated technologies* have been explored in a *case study* by Geng et al. [4]. Other gait biometric integrated techniques that have been proposed include: a fusion of the *gait* and *ear*-biometric to manage low quality samples where the face may be distorted [9], a fusion of *speech* and gait models for human emotion recognition [8], and a fusion of *body geometry* and gait [12], to name a few.

References

1. Barry, Tom. 2002, April. Atlanta Business Chronicle: http://www.bizjournals.com/atlanta/stories/2002/04/22/focus4.html?s=print.
2. Boulgouris, Nikolaos V., Dimitrios Hatzinakos, and Konstantinos N. Plataniotis. 2005. Gait recognition: A challenging signal processing technology for biometric identification. *IEEE Signal Processing Magazine* 22(6): 78-90.
3. Cattin, Philippe C. 2002. Biometric authentication system using human gait. Ph.D. Thesis. Swiss Federal Institute of Technology: Zurich, Switzerland.
4. Geng, Xin, Kate Smith-Miles, Liang Wang, Ming Li, and Qiang Wu. 2010. Context-aware fusion: A case study on fusion of gait and face for human identification in video. *Pattern Recognition* 43(10): 3660–3673.
5. Hossain, Emdad. 2014. Investigating adaptive multi-modal approaches for person identity verification based on face and gait fusion. Ph.D. Dissertation, University of Canberra.
6. HRL Laboratories, LLC. (2015, September). https://www.hrl.com/news/2015/0930/.
7. Kale, Amit, Amit K. Roy Chowdhury, and Ramalingam Chellappa. 2004. Fusion of gait and face for human identification. *IEEE International Conference on Acoustics, Speech, and Signal Processing*, 901–904. Montreal.
8. Lim, Angelica, and Hiroshi G. Okuno. 2012. Using speech data to recognize emotion in human gait. *IEEE/RSJ HBU Workshop* 7559: 52–64.
9. Mark. S. Nixon, Imed Bouchrika, Banafshe Arbab-Zavar, and John N. Carter. 2010. On use of biometrics in forensics: Gait and ear. *European Signal Processing Conference,* 1655–1659. Aalborg.

10. Panchumarthy, Ravi, Ravi Subramanian, and Sudeep Sarkar. 2012. Biometric evaluation in the cloud: A case study with humanID gait challenge. *IEEE/ACM 5th International Conference on Utility and Cloud Computing*, 219–222. Chicago.
11. *Plantiga*. http://www.plantiga.com.
12. Putz-Leszczynska, Joanna, and Miroslaw Granacki. 2014. Gait biometrics with a Microsoft Kinect sensor. *2014 International Carnahan Conference on Security Technology (ICCST)*, 1–5. Rome.
13. Sarkar, Sudeep, et al. 2005. The HumanID gait challenge problem: Data sets, performance, and analysis. *IEEE Transactions on Pattern Analysis and Machine Intelligence* 27(2): 162–177.
14. ViTRAK Systems Inc.: http://stepscan.com.
15. Zikto: http://zikto.com/1/w/.

Chapter 11
Conclusion and Remarks

The study of *behavioral biometrics* has revealed a number of powerful new person-distinguishing characteristics, some of which have the potential to be less intrusive and more fraud-resistant when compared to other security mechanisms such as physical biometrics. However, the complicated data samples that are often associated with the behavioral biometrics impose the need for devising a new generation of processing and classification techniques; that is, techniques that are able to identify the key traits while not being significantly influenced by the *intra*-person variability. In this book, a comprehensive analysis of such techniques has been presented along with the sources of variability, with the intended purpose of performing person recognition via the behavioral biometric known as footstep *ground reaction force* (GRF), which is a form of *gait biometric*. Through our work, two novel *machine learning*-based normalization techniques were proposed, which support two assertions related to the effects of *shoe type* and *stepping speed* on the GRF recognition performance. Furthermore, we have compared a number of *feature extractor*, *normalizer*, and *classifier* configurations that had never before been cross-examined with respect to GRF-based gait recognition. In this concluding chapter, we summarize these findings and we focus on ways in which the work presented in this book may be improved upon by future research. The *trends* and *challenges* that we believe will shape the future development of gait biometrics are also explored. Finally, we discuss some of the most recent *advances* in gait biometric recognition and where they may fit with respect to the work presented in this book.

11.1 Conclusion

This book has made several significant contributions to the study of gait-based person recognition and to the wider field of machine learning. The core contributing content of the book was divided across a two-phase *demonstrative experiment*. In the first phase of this experiment (Chaps. 4–6), we have presented the methodology behind various *machine learning*-based techniques while also demonstrating how each of these methods could be optimally tuned to the GRF biometric; this phase

© Springer International Publishing Switzerland 2016
J.E. Mason et al., *Machine Learning Techniques for Gait Biometric Recognition*,
DOI 10.1007/978-3-319-29088-1_11

has been referred to as *component optimization*. Next, in Chaps. 7–9, we have presented an evaluation of each of the described GRF-based gait recognition techniques on a previously unused dataset. This is referred to as the *evaluation* phase of our experiment. With respect to the GRF-based gait recognition, the work presented in this book backed the assertions that *intra*-person shoe type variations have a negative impact on the recognition performance and that *normalizing* for the stepping speed will have a positive impact on the recognition performance. With respect to *machine learning*, an in-depth theoretical overview of many existing machine learning-based techniques has been provided, along with two novel data preprocessing techniques, namely the *localized least squares regression* (LLSR) normalizer and the *localized least squares regression with dynamic time warping* (LLSRDTW) normalizer.

Concerning the first GRF recognition assertion, to the best of our knowledge, no prior GRF-based gait recognition study has performed a thorough analysis of the effect of *shoe type* on the recognition performance. To address this research gap, our evaluation dataset was divided into three different testing sets: the first testing set containing the samples that always had the *same* shoe type as the training set, the second testing set containing only the samples with a *different* shoe type from the training set, and the third testing set containing a *mix* of samples, where some were from the same shoe type and others were from different shoe types. After applying five different classifiers to evaluate these subsets we discovered an average decrease in the recognition performance of about 50 % when the shoes that were tested differed from those on which a classifier was trained.

When evaluating our second assertion, we could not find any previous supporting GRF-based gait recognition studies; nor could we find much information in the literature dealing with effective statistical/machine learning preprocessing techniques that account for *temporal* variation between data samples. In our work, we postulated that models could be derived to map the relationship between the *step duration* and the *amplitude* at various localities within the GRF signature. To prove this assertion, we developed a new normalization technique called LLSR; this technique generated its models via aligning the training data at each data point in relation to the total step duration and then acquiring the resulting *least squares* regression relationships. We later used the *dynamic time warping* (DTW) technique to better align the samples prior to acquiring the regression models, yielding another normalization technique referred to as LLSRDTW. To perform the actual *normalization* on our dataset, we then assigned a common step duration to each sample, and used the derived models to adjust the sample GRF signatures accordingly. In our evaluation experiments, we compared the non-normalized GRF recognition performance against the performance achieved by using our two novel normalization techniques and a third technique known as *linear time normalization* (LTN). Our findings revealed that each step duration normalizer resulted in an improvement in the recognition performance for almost every feature space, with the exception of the *geometric* feature space. The largest improvement came from our new LLSRDTW normalizer, with a 14–15 % increase in the GRF recognition performance compared to its non-normalized equivalent over our two best feature spaces, namely the holistic and wavelet spaces.

Aside from supporting the two aforementioned assertions, this book also made several other contributions to the study of GRF-based gait recognition. To the best of our knowledge, our examined sample space was larger than the sample spaces used in all previous GRF-based gait recognition researches, with about 12,800 points per sample covering 8 different GRF signals. Furthermore, this book is the first attempt, to our knowledge, at investigating GRF recognition using the *least square probabilistic classification* (LSPC) classifier. It is also the first time, for the purpose of GRF recognition, that (1) the *wavelet packet* feature extraction technique has been compared against the *spectral* and *holistic* techniques, (2) the *holistic* feature extraction technique has been compared against the *spectral* technique, and (3) the *multilayer perceptron* (MLP) classifier has been compared against the *linear discriminant analysis* (LDA) classifier. Through our experiments, we have demonstrated that the wavelet packet feature extractor is superior to all other tested feature extractors, with an average EER of about 2.6 % and the best EER of about 1.6 %. Moreover, in our evaluation tests, we found that the LDA classifier outperforms other tested classifiers. It performed roughly 19 % better than the next best classifier on average; however, some performance concerns were raised with regard to its applicability in a practical setting.

The research presented in this book is tied together with the analysis of a number of powerful *preprocessing* and *classification* techniques, providing a foundation upon which the future work can be built. While the objectives of this book were achievable over a relatively limited dataset of only 10 different subjects, a future analysis would benefit greatly from the use of a more realistic dataset. Ideally, the future dataset would contain over 100 different subjects with varying *footwear* and with the steps collected at varying *walking speeds*. During testing, this dataset should be divided into *enrolled* subjects and previously unseen *imposter* subjects. In analyzing such a dataset, it is expected there would be a more accurate measure of the performance in a real-world setting. It is also expected that many of our *machine learning*-based preprocessors and classifiers will benefit from the inclusion of more samples during the training phase. However, a larger dataset may expose some of the potential flaws in our better performing biometric system configurations, namely the computational inefficiencies. A future research project might be to perform a thorough analysis of the computational growth rate introduced by each configuration in order to achieve a greater understanding of the underlying *performance-efficiency* trade-offs. We also believe that the study of the footstep GRF-based gait biometric can benefit greatly from a similar assessment of the impact of the training/testing feet (*left* vs. *right*) on the recognition performance.

In terms of improving the recognition performance established in this book, our experimental analysis made several suggestions. Many of our existing classifiers likely have room for improvement through their direct use in the optimization of the preprocessors. In our work, all the preprocessors were optimized exclusively using the *K-nearest neighbor* (KNN) classifier. We believe that any future analysis will benefit from the inclusion of *alternative* feature extractors and classifiers. New classifiers to be tested might include the *hidden Markov model* (HMM) classifier [5], the *Max-Margin Markov Network* (M^3-net) classifier [8], and/or different variants of the classifiers examined in this book, such as the *regularized-LDA* (RLDA) classifier [4]. New feature extractors to be tested might include the *partial least squares* (PLS) [12], the *kernel principal component analysis* (KPCA) [11], and the *generalized PCA* (GPCA) [3] feature extractors, to name a few. In our experimental analysis there was also a discussion with regard to how the DTW could be used along with the *center star* approximation algorithm as a *stand-alone* data preprocessor, to be run prior to any feature extraction. Furthermore, we believe that our GRF-based gait recognition performance could be improved by using strategies that allow for the *rejection* of undesirable samples from the training data and the *fusion* of the results obtained over multiple different feature spaces.

In addition to furthering our understanding of the GRF-based gait recognition domain, much of the work presented in this book could also be applicable to a far wider field of classification problems. This is particularly true for the machine learning-based techniques that were used, which we believe could be applied to a number of other domains with little-to-no domain knowledge required. In the future work, the technologies presented in this book and their suggested improvements could be applied to other application domains such as speech recognition [7]. It is our hope that this book can become something of a template when faced with a classification problem in an unfamiliar domain.

11.2   Future Perspectives

Taking a step back from the particulars of the demonstrated GRF-based gait recognition system and turning to the *gait biometric* in more general terms, the work presented in this book, particularly in Chaps. 2 and 10, provides some perspective on the utility and challenges faced in bringing such a concept to a real-world setting. Many of the potential setbacks that might hold back a gait-recognition system are often nontechnical in nature. Moreover, changes in the underlying technology could have significant implications on the future procurement of such systems. Consequently, prior to dedicating further research toward the gait biometric, it is important that we understand how this biometric is *evolving* and in which areas it may be most valuable. The following two subsections revisit the gait biometric from a higher level perspective and expand upon some of the *challenges* and *trends* that may be shaping its future.

11.2.1   Gait Biometric Trends and Challenges

The interest in gait biometric recognition technology has been motivated by the increasing demand for *cost-effective* automated recognition systems for monitoring applications and visual surveillance, and the increased *availability* of new image sensors and low cost computing power. Gait biometric recognition methods include: (1) motion-based *state-space* methods [1]—where the human gait motion is constructed as a sequence of static body poses and temporal variation observations with respect to these body poses used to identify it; (2) motion-based *spatiotemporal* methods [6]—where the spatiotemporal distribution of the human gait motion is characterized, leading to human recognition; (3) *model*-based methods—where the human gait motion is modeled before the image features are extracted by the motion trajectories of body parts [2]; and (4) *holistic*-based approaches [9]—where no specific model for the walking human is assumed. The pros and cons of these approaches are discussed in [2, 9]. In the sequel, we focus on a few challenges with respect to *technology*, *research*, and *practical deployment*.

Challenges in Technology: Gait recognition technology is still in its infancy in development. To the best of our knowledge, there is no model as of yet that has been developed and is sufficiently accurate and marketable. However, gait technology is growing at a rapid pace, with some key technologies currently under development such as automatic analysis of *video* imagery and surveillance, design of *radar* systems that generate radar records of the gait cycle created by the body parts as a subject walks, gait technology for fingerprint, retina scans and face detection, and gait technology for security purposes, for instance, to screen people from much greater distances than they can today [10], to name a few. It is envisaged that gait recognition technology will be *integrated* with other biometric technologies to generate more efficient methods for user identification and authentication. The potential of gait technology as a marketable technology is full of promise and is currently being explored by various corporations.

Challenges in Research: The goal of gait biometric analysis is to characterize biometric information such as gender, age, identity, and ethnicity based on people's walking patterns. Given several image sequences capturing a human walking, some of the recent research trends in gait biometric recognition have focused on devising some analytical models for studying the *spatial* and *temporal* features, in order to derive *dynamic* and *static* models of the human motion. Other research trends include identifying and extracting more robust features, and improving the accuracy of these analytical models and their applicability in real-world surveillance systems settings.

In gait biometric recognition, much effort has been devoted to identifying the factors that may negatively influence the *accuracy* of such systems, and methods for more *cost-effective* recognition accuracy have been proposed. But, fewer research works have been devoted to ensuring the robustness of gait systems against threats such as passive and dynamic *imposter* attacks, *phishing* attacks, and *spoofing* attacks, to name a few. To make a gait biometric application successful, considerations such as users' fear of unfamiliar technology, invasion of *privacy*, misconception about the purpose and usage of gait biometric information, and use of gait human motions profiles for *unauthorized* activities, are among the main concerns when deploying gait recognition technologies. It is also widely believed that one of the key concerns of *any* biometric system, including gait biometric recognition system, is the violation of a citizen's right to *anonymity* since his/her privacy can be invaded. For instance, access to gait biometric information by an unauthorized user can be exploited for unlawful purposes. Also, the study of gait *patterns* such as psychological condition of a human, fatigue, etc., may constitute a serious threat to the privacy of a person's *medical* information. In this respect, one of the most practical solutions is to design a method to only retain the gait sequence information that is useful for recognition purpose, rather than considering the entire volume of available information.

Challenges for Practical Deployment: Most existing theoretical approaches for gait recognition have been proven to achieve remarkable performances in terms of accuracy rate [13]. However, due to factors affecting the walking pattern such as *speed*, walking *surface*, and various other factors such as *physiology* and *psychological* condition, which have not yet been widely studied, and which can cause some changes in the natural gait, gait recognition for *personal identification* is yet to be widely deployed for practical applications. Such applications may include video surveillance, forensics investigation, medicine, criminal identification based on video sequences, and user authentication on mobile devices, to name a few. Indeed, the design and suitability of gait biometric technology for person identification depends on the application *requirements*, which for instance, can be specified in terms of *throughput*, system *security*, user *acceptance*, identification *accuracy*, return on *investment*, robustness, etc. For the next generation of gait biometric recognition systems, several issues are yet to be addressed to improve the recognition accuracy. These challenges include handling incomplete or poor *quality* of data, *scaling* the system in terms of accommodating a large number of users, ensuring *interoperability*, protecting the user's *privacy and system integrity*, and minimizing the system *cost*, to name a few.

11.2.2 Advances in Biometric Gait Recognition

Due to the cyclic combination and coordinated nature of the motion that characterizes the gait biometric, several advances in gait recognition can be highlighted, a few of which are as follows. As a source of entertainment, *athletics* may derive value if the gait biometric is utilized in the sense that human motion can be evaluated to predict the athletic potential and evaluate the training of an individual. In computer graphics and the gaming industry, *motion capture* plays a key role. Therefore, advances in human motion analysis can significantly improve *markerless* systems, which are expected to become the standard norm for motion capture. In *healthcare*, the current practice is to diagnose and monitor the treatments using human observations. Gait analysis has led to a way to improve the diagnosis of gait-related disorders, as well as monitoring of the treatment of these disorders. Consequently, while the use of gait for person verification or identification explored in this book may face hurdles in achieving a wide acceptance in the security community, the study of gait recognition offers a number of alternative commercially applicable avenues for further exploration that have only recently become technically feasible.

References

1. BenAbdelkader, Chiraz, Ross Cutler, Harsh Nanda, and Larry Davis. 2001. EigenGait: Motion-based recognition of people using image self-similarity. In *Third international conference on audio and video based biometric person authentication*, 284–294, Halmstad.
2. Boulgouris, Nikolaos V., Dimitrios Hatzinakos, and Konstantinos N Plataniotis. 2005. Gait recognition: A challenging signal processing technology for biometric identification. *IEEE Signal Processing Magazine* 22(6): 78–90.
3. Cattin, Philippe C. 2002. Biometric authentication system using human gait, Ph.D. Thesis 2002. Switzerland: Swiss Federal Institute of Technology.
4. Friedman, Jerome H., 1989. Regularized discriminant analysis. *Journal of the American Statistical Association* 84(405): 165–175.
5. Ghahramani, Zoubin, 2001. An introduction to hidden markov models and bayesian networks. *International Journal of Pattern Recognition and Artificial Intelligence* 15(1): 9–42.
6. Han, Ju, and Bir Bhanu. 2006. Individual recognition using gait energy image. *IEEE Transactions on Pattern Analysis and Machine Intelligence* 28(2): 316–322.
7. Juang, Biing-Hwang, and Lawrence R. Rabiner. 2005. *Automatic Speech Recognition—A Brief History of the Technology Development*, 2nd ed. Elsevier: Encyclopedia of Language and Linguistics. Available at: http://www.ece.ucsb.edu/Faculty/Rabiner/ece259/Reprints/354_LALI-ASRHistory-final-10-8.pdf.
8. Julien, Simon-Lacoste, 2003. *Combining SVM with graphical models for supervised classification: An introduction to Max-Margin Markov Networks*. Berkeley, CA, USA: University of California, Berkeley. Project report 2003.
9. Masupha, Lerato, Tranos Zuva, and Salemen Ngwira. 2015. A review of gait recognition techniques and their challenges. In *Third international conference on digital information processing, e-business and cloud computing*, 63–69, Reduit.
10. Sarkar, Sudeep, and Zongyi Liu. 2008. Gait recognition. In *Handbook of Biometrics*, eds. Anil K. Jain, Patrick Flynn, and Arun A. Ross, Ch. 6, 109–129. New York: Springer.
11. Schölkopf, Bernhard, Alexander Smola, and Klaus-Robert Müller. 1997. Kernel principal component analysis. In *Artificial neural networks—ICANN'97*, 583–588. Berlin: Springer.
12. Schwartz, William Robson, Aniruddha Kembhavi, David Harwood, and Larry S. Davis. 2009. Human detection using partial least squares analysis. In *IEEE 12th international conference on computer vision*, 24–31, Kyoto.
13. Sudha, Livingston R., and R. Bhavani. 2012. Gait based gender identification using statistical pattern classifiers. *International Journal of Computer Applications* 40(8): 0975–8887.

Appendix
Experiment Code Library

The code library developed to evaluate the various biometric system configurations and machine learning techniques described in this book is made available at the link below. Use the **13 digit ISBN**, located at the beginning of the book, to gain access.

http://www.uvic.ca/engineering/ece/isot/books/index.php

On this web page you will find two versions of the library, one written for **Java** developers and the other for **.NET** developers. Future versions may also feature a user interface for researchers with limited programming experience. This code was initially developed to discover and classify biometric features exclusively within our 8 signal Ground Reaction Force domain, but has since been modified to fit more general biometric recognition problems. Outputs from this library can optionally be written to **CSV** files for easy analysis in Microsoft Excel.

The *highlights* of this library include the following:

- **Biometric Evaluation**

  - K-fold cross-validation
  - Equal error rates
  - False acceptance rates
  - False rejection rates

- **Data Formatters**

  - Data amplitude rescaling
  - Area-based sample length rescaling
  - Point-based sample length rescaling

- **Feature Extractors**

  - Geometric and optimal geometric extraction
  - Holistic extraction (*PCA*)
  - Spectral extraction (*periodogram, magnitude spectra*)
  - Wavelet extraction (*WPD, Fuzzy c-means*)

© Springer International Publishing Switzerland 2016
J.E. Mason et al., *Machine Learning Techniques for Gait Biometric Recognition*,
DOI 10.1007/978-3-319-29088-1

- **Normalizers**
  - L-type (L^1, L^2, L^∞)
  - Score normalization
  - Linear time normalization
  - Localized least squares regression
  - Localized least squares regression with dynamic time warping (*Center Star algorithm, DTW*)

- **Classifiers**
  - K-nearest neighbors
  - Multilayer perceptron
  - Support vector machine
  - Linear discriminant analysis (*ULDA, KUDA*)
  - Least squares probabilistic classification
  - Parameter optimization

Index

A
Accord.NET, 66
Akaike information criterion, 13
Amplitude warping function, 97, 107
Analysis of covariance, 94
  covariate, 94
ANCOVA. *See* Analysis of covariance
Artificial neural network. *See* Neural network
Authentication
  continuous, 27, 204
  gait, 22
  static, 204

B
Backpropagation, 118, 119
Bayes' rule, 46, 133
Biometric evaluation
  data quality, 157
  security, 161
  usability, 158
Biometric system, 16, 157
  bias, 120, 157, 171, 175, 176
  challenge phase, 19, 161
  components, 21, 157
  enrollment phase, 19, 161
  experimental, 163
  modes of operation, 204
Biometrics
  at a distance, 2
  behavioral, 1, 2, 209
  physical, 1
  recognition, 1
  spoofing, 1, 2
Boeing, 206
Bootstrapping, 13

C
Case study, 206, 207
Center star, 100, 102–106, 199
Character, 157
Classification, 45, 111
  boundaries, 111, 116, 122
  discriminative, 46, 132, 198
  generative, 46, 132, 198
  similarity-based, 16
Code library
  .NET, 217
  Java, 217
Coiflet, 80, 197
Covariance matrix, 65, 133
Covariate factors, 27, 42, 206

D
DARPA, 2, 206
Data smoothing, 59
Dataset, 10
  development, 21, 47, 49, 167
  evaluation, 21, 47, 49, 167
  experimental, 167
Daubechies, 80
DET. *See* Detection error tradeoff
Detection and identification rate, 18, 158
Detection error tradeoff, 17, 18, 178
Dimensionality reduction, 16, 85
  Fisher, 133, 134, 136
  geometric, 62
  holistic, 65, 70
  spectral, 72, 77
  supervised, 199
  wavelet, 80, 82, 84
Discrete fourier transform, 74

© Springer International Publishing Switzerland 2016
J.E. Mason et al., *Machine Learning Techniques for Gait Biometric Recognition*,
DOI 10.1007/978-3-319-29088-1

Normalization (*cont.*)
  L^2, 90
  L^∞, 90, 102
  linear time. *See* Linear time normalization
  localized least squares regression. *See* Localized least squares regression
  localized least squares regression with dynamic time warping. *See* Localized least squares regression with dynamic time warping
  score, 93
  step duration, 179, 192, 210

O
Optimization parameter, 111
  hidden nodes, 120
  K, 114
  kernel, 129, 141, 149
  learning rate, 120
  momentum, 120
  regularization, 127, 147
Optimization technique, 15
Overfitting, 53, 142

P
Partial least squares, 199
PCA. *See* Principal component analysis
Periodogram, 74
Phase warping function, 107
Plantiga, 37, 206
Pooled within-group regression, 96
Posterior probability, 15, 153, 164
  K-Nearest Neighbors, 113
  least squares probabilistic classification, 145
  linear discriminant analysis, 133, 141
  multilayer perceptron, 120
  support vector machine, 129
Power spectral density, 72, 74
Pre-processing, 21, 110, 196
  data abstraction, 11
  data aggregation, 11
  data cleaning, 11
  data normalization, 11
  data reduction, 11
Principal component, 65
Principal component analysis, 43, 65, 67, 134, 136
  generalized, 199
  kernel, 199
  projection, 66

Probability density function, 113, 136
PSD. *See* Power spectral density

Q
Quadratic discriminant analysis, 133

R
Random subsampling cross-validation, 13
Receiver operating characteristics, 18
Regression model, 94
Rescaling, 11, 114, 120
Residuals, 95

S
Sakoe-Chiba Band, 101, 199
Scatter matrix
  between-class, 134
  within-class, 134
Sensitivity threshold, 58
Sigmoid function, 117
Singular value decomposition, 67, 137
  Reduced, 137
Small sample size, 133
Social engineering, 23
Spectral
  leakage, 74
  magnitude. *See* Magnitude spectra
Spectral features, 44, 72
Standard score. *See* Z-score
Standardization
  area-based, 67
  point-based, 67
Step duration, 94
Support vector machine, 16, 46, 122
  primal optimization problem, 123, 126
  slack variable, 126
  soft margin, 126
Surveillance, 17, 24, 203, 213
SVD. *See* Singular value decomposition
SVM. *See* Support vector machine

T
Time-frequency domain, 78
Trapezoidal rule, 61
Triangle approximation, 60

U
ULDA. *See* Uncorrelated linear discriminant analysis
Uncorrelated linear discriminant analysis, 133, 136